# World Atlas Student Workbook Featuring Maps from the Rand McNally Goode's World Atlas, 21st Edition

Eugene J. Palka
*United States Military Academy*

Jon C. Malinowski
*United States Military Academy*

JOHN WILEY & SONS, INC.

"The opinions expressed are solely those of the authors and do not represent those of the Department of Army, the Department of Defense, or the United States Military Academy."



To order books or for customer service, please call 1-800-CALL-WILEY (225-5945).

ISBN 0-471-70691-4

Printed in the United States of America

10 9 8 7 6 5 4 3 2 1

Printed and bound by Courier Kendallville, Inc.

# *Table of Contents*

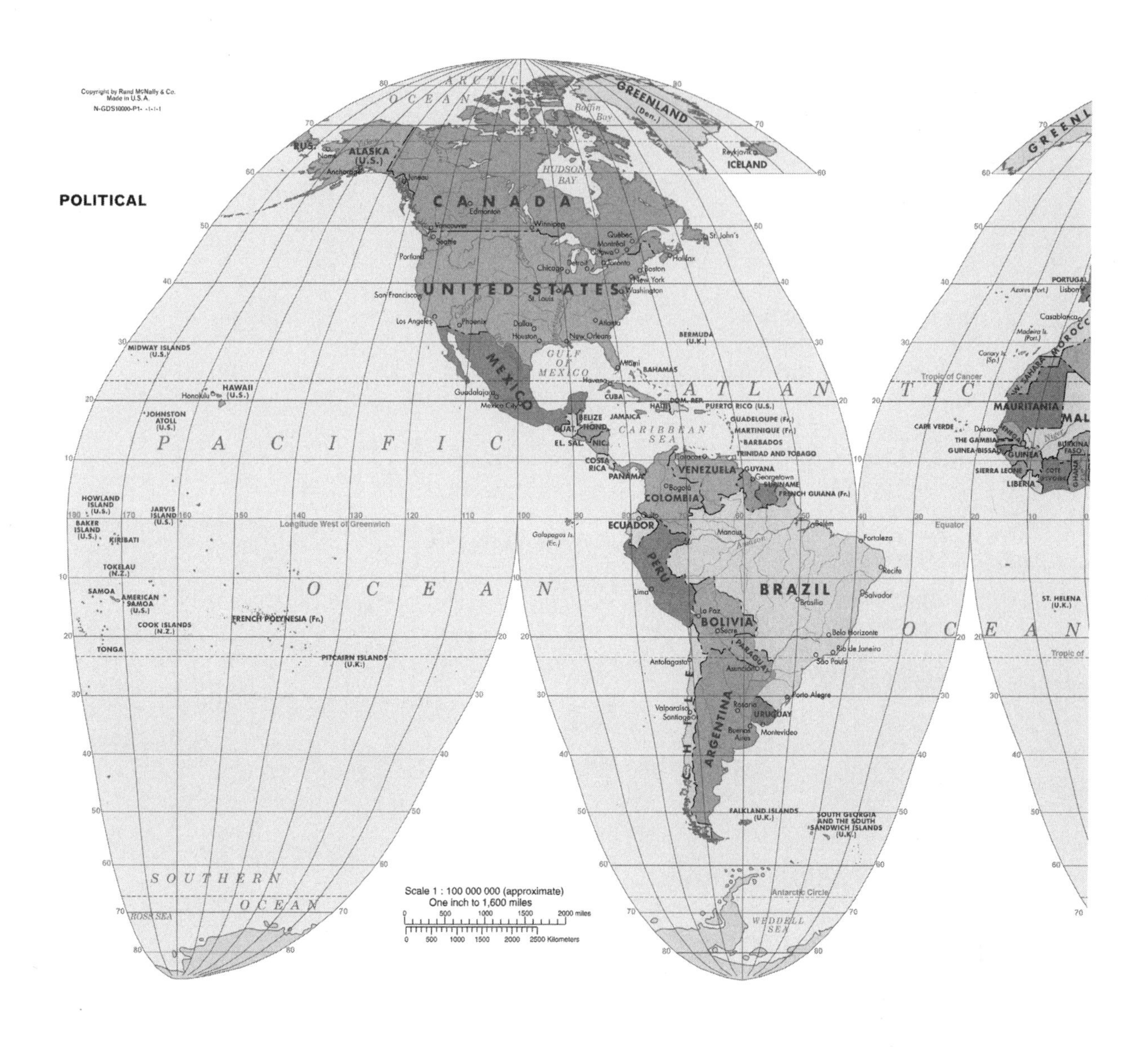

Copyright by Rand McNally & Co.
Made in U.S.A.
POLITICAL
ARCTIC OCEAN
GREENLAND (Den.)
Baffin Bay
ICELAND
Reykjavik
ALASKA (U.S.)
RUS.
Anchorage
Juneau
HUDSON BAY
CANADA
Edmonton
Vancouver
Seattle
Winnipeg
Québec
Montréal
Ottawa
Toronto
St. John's
Halifax
Portland
Chicago
Detroit
Boston
New York
Washington
San Francisco
UNITED STATES
St. Louis
Los Angeles
Phoenix
Dallas
Atlanta
Houston
New Orleans
GULF OF MEXICO
Miami
BAHAMAS
BERMUDA (U.K.)
MIDWAY ISLANDS (U.S.)
HAWAII (U.S.)
Honolulu
MEXICO
Havana
CUBA
Guadalajara
Mexico City
HAITI
DOM. REP.
PUERTO RICO (U.S.)
GUADELOUPE (Fr.)
MARTINIQUE (Fr.)
BARBADOS
TRINIDAD AND TOBAGO
BELIZE
JAMAICA
GUAT.
HOND.
EL SAL.
NIC.
CARIBBEAN SEA
COSTA RICA
PANAMA
Caracas
VENEZUELA
GUYANA
Georgetown
SURINAME
FRENCH GUIANA (Fr.)
Bogotá
COLOMBIA
JOHNSTON ATOLL (U.S.)
PACIFIC OCEAN
ATLANTIC OCEAN
HOWLAND ISLAND (U.S.)
BAKER ISLAND (U.S.)
JARVIS ISLAND (U.S.)
KIRIBATI
Longitude West of Greenwich
Galapagos Is. (Ec.)
ECUADOR
Manaus
Belém
Fortaleza
Recife
TOKELAU (N.Z.)
SAMOA
AMERICAN SAMOA (U.S.)
COOK ISLANDS (N.Z.)
TONGA
FRENCH POLYNESIA (Fr.)
PITCAIRN ISLANDS (U.K.)
PERU
Lima
BRAZIL
Brasília
Salvador
La Paz
BOLIVIA
Sucre
Belo Horizonte
Rio de Janeiro
São Paulo
Antofagasta
PARAGUAY
Asunción
Porto Alegre
Valparaíso
Santiago
Rosario
URUGUAY
Buenos Aires
Montevideo
CHILE
ARGENTINA
FALKLAND ISLANDS (U.K.)
SOUTH GEORGIA AND THE SOUTH SANDWICH ISLANDS (U.K.)
SOUTHERN OCEAN
ROSS SEA
Antarctic Circle
WEDDELL SEA
Scale 1 : 100 000 000 (approximate)
One inch to 1,600 miles
0 500 1000 1500 2000 miles
0 500 1000 1500 2000 2500 Kilometers
PORTUGAL
Lisbon
Azores (Port.)
Casablanca
Madeira Is. (Port.)
MOROCCO
Canary Is. (Sp.)
W. SAHARA
Tropic of Cancer
MAURITANIA
MALI
CAPE VERDE
Dakar
SENEGAL
THE GAMBIA
GUINEA-BISSAU
GUINEA
BURKINA FASO
SIERRA LEONE
LIBERIA
Equator
ST. HELENA (U.K.)
Tropic of Capricorn


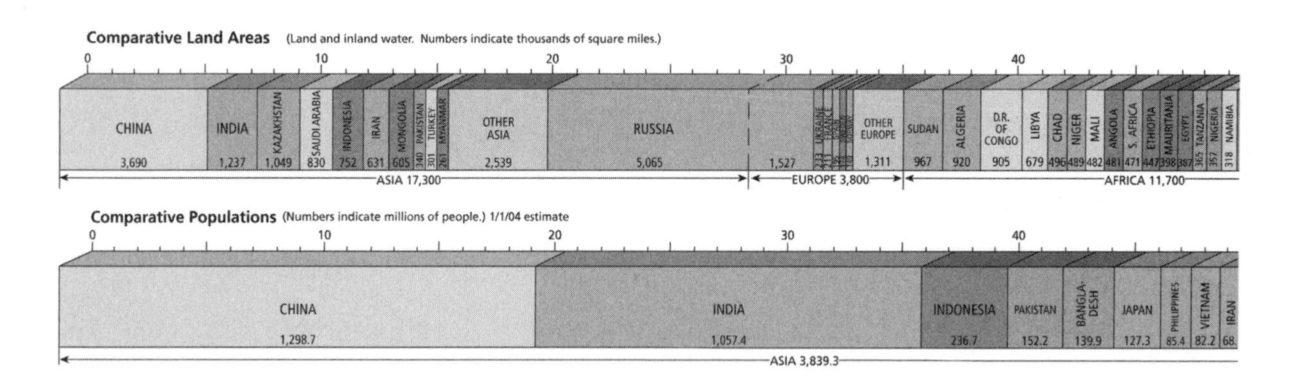

Comparative Land Areas (Land and inland water. Numbers indicate thousands of square miles.)
0 10 20 30 40
CHINA 3,690
INDIA 1,237
KAZAKHSTAN 1,049
SAUDI ARABIA 830
INDONESIA 752
IRAN 631
MONGOLIA 605
PAKISTAN 340
TURKEY 301
MYANMAR 261
OTHER ASIA 2,539
RUSSIA 5,065
1,527
OTHER EUROPE 1,311
SUDAN 967
ALGERIA 920
D.R. OF CONGO 905
LIBYA 679
CHAD 496
NIGER 489
MALI 482
ANGOLA 481
S. AFRICA 471
ETHIOPIA 447
MAURITANIA 398
EGYPT 387
TANZANIA 365
NIGERIA 357
NAMIBIA 318
ASIA 17,300
EUROPE 3,800
AFRICA 11,700
Comparative Populations (Numbers indicate millions of people.) 1/1/04 estimate
0 10 20 30 40
CHINA 1,298.7
INDIA 1,057.4
INDONESIA 236.7
PAKISTAN 152.2
BANGLADESH 139.9
JAPAN 127.3
PHILIPPINES 85.4
VIETNAM 82.2
IRAN 68.
ASIA 3,839.3

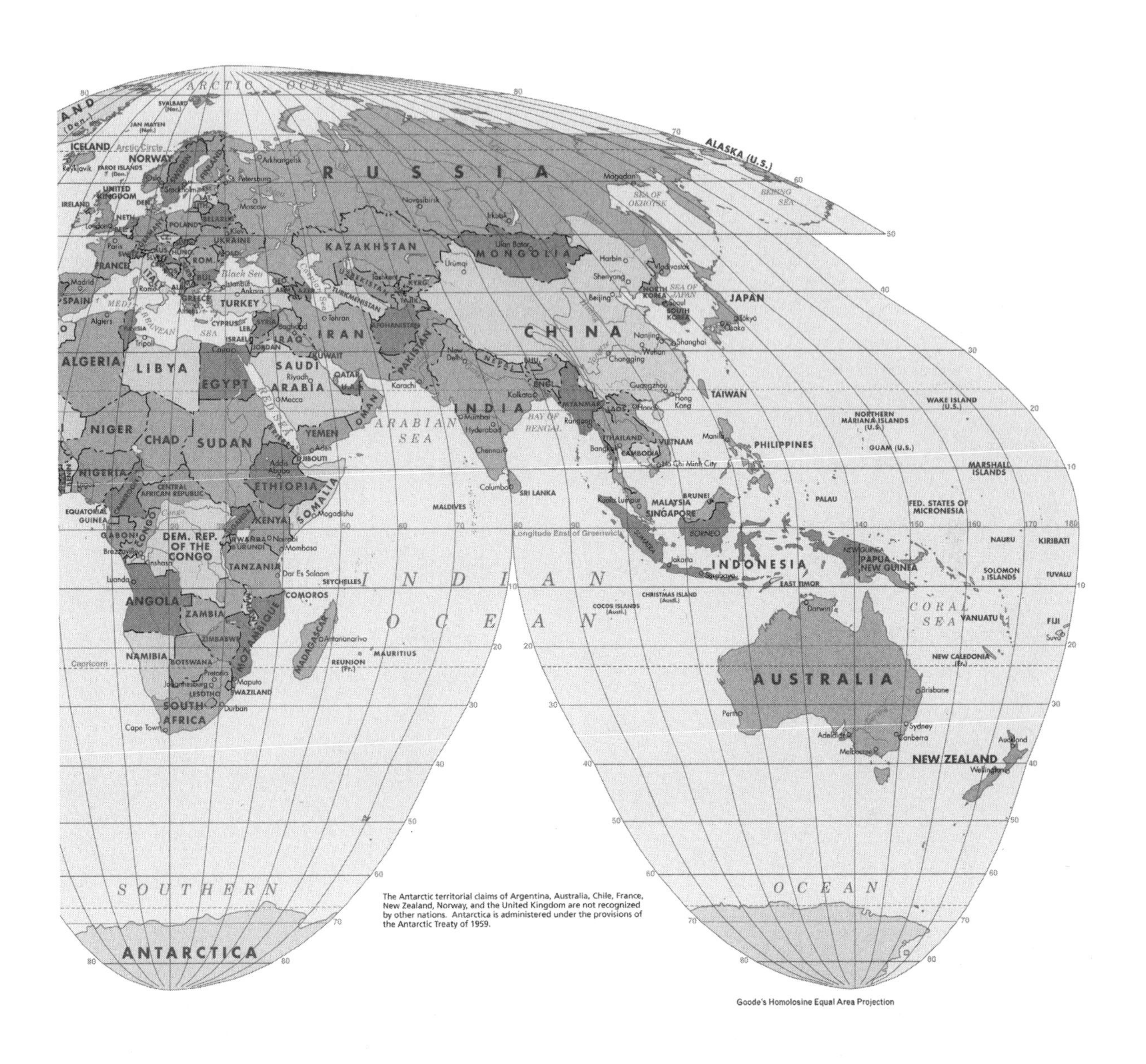
ARCTIC OCEAN
SVALBARD (Nor.)
JAN MAYEN (Nor.)
ICELAND
Reykjavík
FAROE ISLANDS (Den.)
NORWAY
UNITED KINGDOM
IRELAND
London
Paris
FRANCE
Madrid
SPAIN
Rome
Algiers
Arkhangelsk
St. Petersburg
Moscow
POLAND
UKRAINE
ROM.
Black Sea
Istanbul
Ankara
TURKEY
GREECE
CYPRUS
Athens
Tripoli
ISRAEL
JORDAN
SYRIA
IRAQ
Baghdad
Tehran
IRAN
KUWAIT
QATAR
Riyadh
SAUDI ARABIA
Mecca
RUSSIA
Novosibirsk
Magadan
ALASKA (U.S.)
SEA OF OKHOTSK
BERING SEA
KAZAKHSTAN
Tashkent
TURKMENISTAN
AFGHANISTAN
PAKISTAN
Karachi
Ürümqi
MONGOLIA
Ulan Bator
Harbin
Vladivostok
Shenyang
Beijing
NORTH KOREA
SEA OF JAPAN
Seoul
SOUTH KOREA
JAPAN
Tokyo
Osaka
CHINA
Nanjing
Shanghai
Wuhan
Chongqing
Guangzhou
Hong Kong
TAIWAN
New Delhi
NEPAL
Kolkata
INDIA
Mumbai
Hyderabad
Chennai
BAY OF BENGAL
ARABIAN SEA
Colombo
SRI LANKA
MALDIVES
Yangon
THAILAND
LAOS
Hanoi
Bangkok
CAMBODIA
VIETNAM
Ho Chi Minh City
Manila
PHILIPPINES
Kuala Lumpur
MALAYSIA
SINGAPORE
BRUNEI
BORNEO
INDONESIA
Jakarta
EAST TIMOR
PAPUA NEW GUINEA
PALAU
GUAM (U.S.)
NORTHERN MARIANA ISLANDS (U.S.)
WAKE ISLAND (U.S.)
MARSHALL ISLANDS
FED. STATES OF MICRONESIA
NAURU
KIRIBATI
SOLOMON ISLANDS
TUVALU
VANUATU
FIJI
Suva
NEW CALEDONIA (Fr.)
CORAL SEA
Darwin
AUSTRALIA
Brisbane
Perth
Sydney
Canberra
Adelaide
Melbourne
Auckland
Wellington
NEW ZEALAND
CHRISTMAS ISLAND (Austl.)
COCOS ISLANDS (Austl.)
INDIAN OCEAN
ALGERIA
LIBYA
EGYPT
Cairo
NIGER
CHAD
SUDAN
NIGERIA
YEMEN
Aden
DJIBOUTI
OMAN
ETHIOPIA
SOMALIA
Mogadishu
CENTRAL AFRICAN REPUBLIC
EQUATORIAL GUINEA
GABON
CONGO
DEM. REP. OF THE CONGO
KENYA
Nairobi
RWANDA
BURUNDI
Mombasa
Kinshasa
Luanda
TANZANIA
Dar Es Salaam
SEYCHELLES
COMOROS
ANGOLA
ZAMBIA
ZIMBABWE
MOZAMBIQUE
MADAGASCAR
Antananarivo
MAURITIUS
REUNION (Fr.)
NAMIBIA
BOTSWANA
Pretoria
Johannesburg
Maputo
SWAZILAND
LESOTHO
Durban
SOUTH AFRICA
Cape Town
Capricorn
Longitude East of Greenwich
SOUTHERN OCEAN
ANTARCTICA
The Antarctic territorial claims of Argentina, Australia, Chile, France, New Zealand, Norway, and the United Kingdom are not recognized by other nations. Antarctica is administered under the provisions of the Antarctic Treaty of 1959.
Goode's Homolosine Equal Area Projection

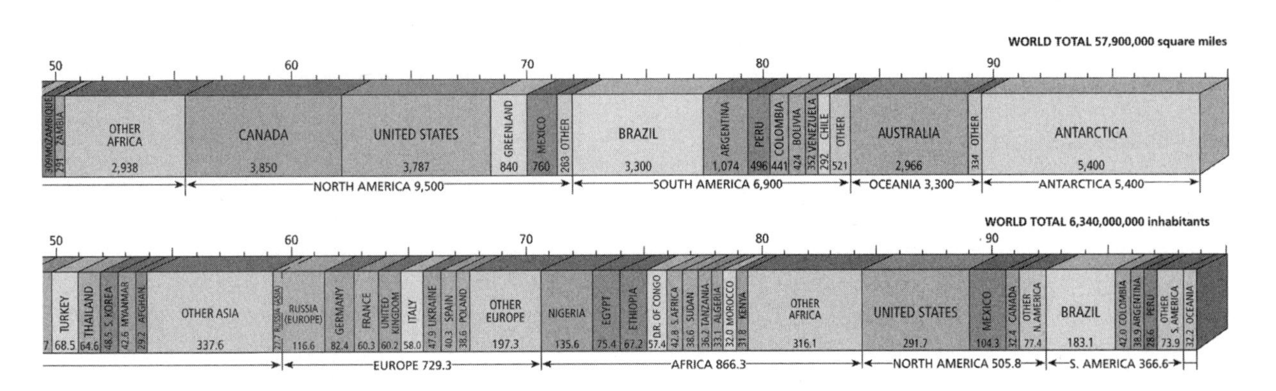
WORLD TOTAL 57,900,000 square miles
50
60
70
80
90
309 MOZAMBIQUE
291 ZAMBIA
OTHER AFRICA 2,938
CANADA 3,850
UNITED STATES 3,787
GREENLAND 840
MEXICO 760
OTHER 263
BRAZIL 3,300
ARGENTINA 1,074
PERU 496
COLOMBIA 441
BOLIVIA 424
VENEZUELA 352
CHILE 292
OTHER 521
AUSTRALIA 2,966
OTHER 334
ANTARCTICA 5,400
NORTH AMERICA 9,500
SOUTH AMERICA 6,900
OCEANIA 3,300
ANTARCTICA 5,400
WORLD TOTAL 6,340,000,000 inhabitants
50
60
70
80
90
TURKEY 68.5
THAILAND 64.6
S. KOREA 48.5
MYANMAR 42.6
AFGHAN. 29.2
OTHER ASIA 337.6
RUSSIA (ASIA) 27.7
RUSSIA (EUROPE) 116.6
GERMANY 82.4
FRANCE 60.3
UNITED KINGDOM 60.2
ITALY 58.0
UKRAINE 47.9
SPAIN 40.3
POLAND 38.6
OTHER EUROPE 197.3
NIGERIA 135.6
EGYPT 75.4
ETHIOPIA 67.2
DR. OF CONGO 57.4
S. AFRICA 42.8
SUDAN 38.6
TANZANIA 36.2
ALGERIA 33.1
MOROCCO 32.0
KENYA 31.8
OTHER AFRICA 316.1
UNITED STATES 291.7
MEXICO 104.3
CANADA 32.4
OTHER N. AMERICA 77.4
BRAZIL 183.1
COLOMBIA 42.0
ARGENTINA 38.9
PERU 28.6
OTHER S. AMERICA 73.9
OCEANIA 32.2
EUROPE 729.3
AFRICA 866.3
NORTH AMERICA 505.8
S. AMERICA 366.6

United States Capitol, Washington, D.C. *PhotoDisc, Inc.*

## *World Political Map*

Name: ______________________ Date: ______________________

See pages **2-3** in Goode's World Atlas (21e) for full color maps.

List a **country** for each of the following letters.

| | |
|---|---|
| A ustrallia | M orocco |
| B razil | N ew Zeland |
| C anada | O man |
| D enmark | P ortugal |
| E thopia | Q atar |
| F rance | R ussia |
| G hana | S yria |
| H olland | T urkey |
| I raq | U kuraine |
| J ordan | V ietnam |
| K enya | W |
| L ibia | Y emen |

List all the countries that border the Equator.

Brazil Gabon Kenya Indonesia
Columbia Congo UGanda Sumatra
Ecuador Dem. Rep. of Congo Somolia Malaysia

List all the countries that have territory above the Arctic Circle.

Canada USA Norway Finland
Russia Greenland Sweeden

What are the southernmost countries on each continent?

South Africa Chila Indonesia Australlia
Panama Greece Antartica

What countries in the world are largely made up of islands?

List ten countries that are landlocked.

Switzerland
Austria
Hungria
Belarus
Armenia
Uzbekistan
Afghanistan
Mongolia

The Prime Meridian (0° longitude) passes through what countries?

England
France
Spain
Algeria
Mali
Burkina Faso
Ghana

Which is in the Northern Hemisphere, the Tropic of Cancer or the Tropic of Capricorn?

Tropic of Cancer

List all the countries that border the Indian Ocean.

India
Sri Lanka
Oman
Tanzania
Yemen
Somalia
Kenya
Mozambique
Madagascar
South Africa
Australlia
Indonesia
Bangladesh
Myanmar
Thailand
Malaysa

List all the countries that border Russia.

Kazakhstan
Mongolia
China
North Korea
Georgia
Azerbaijan
Ukraine
Belarus
Latvia
Estonia
Finland
Norway

Yosemite National Park, California, USA *Corbis Digital Stock*

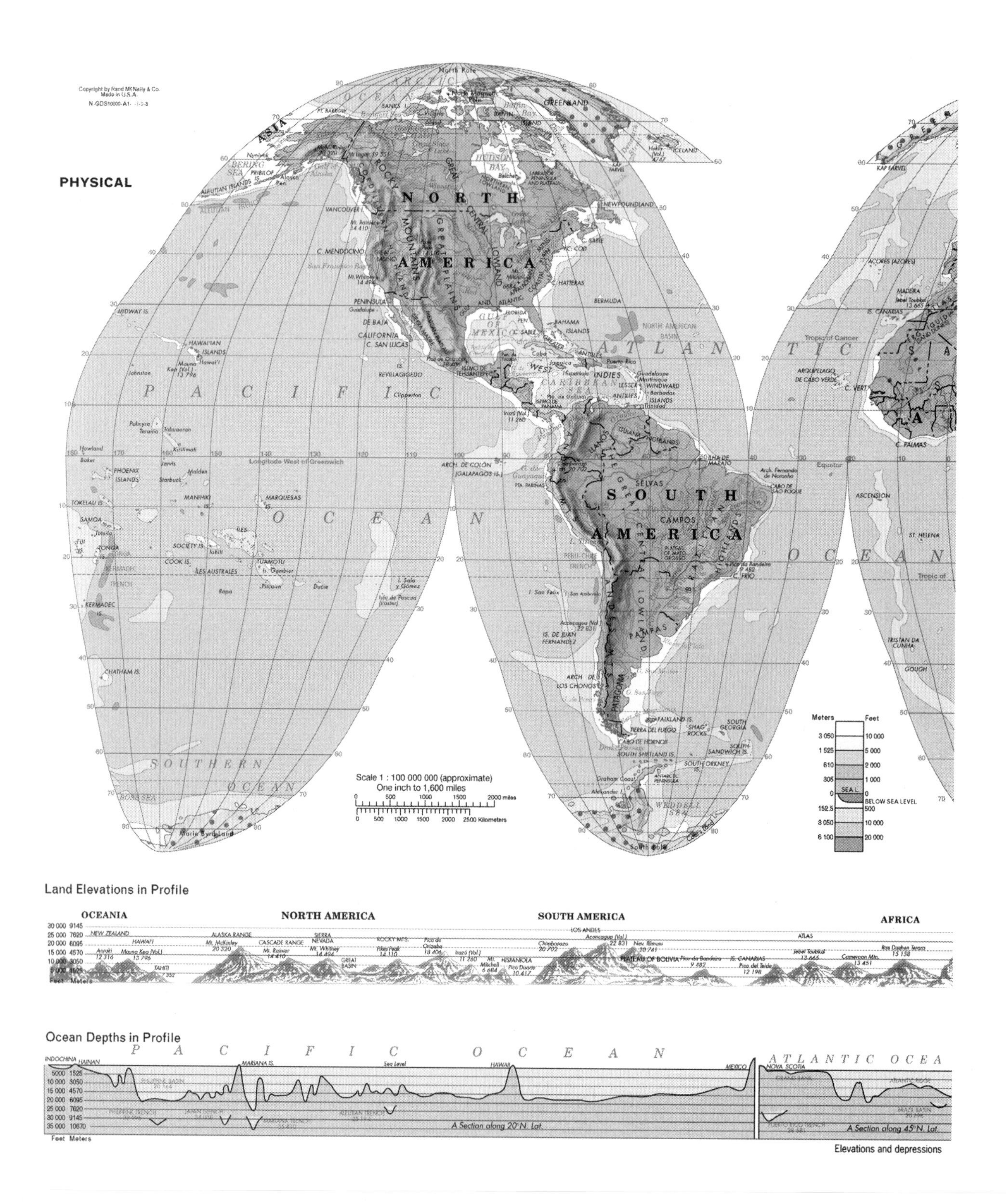
PHYSICAL
Copyright by Rand McNally & Co.
Made in U.S.A.
NORTH AMERICA
SOUTH AMERICA
PACIFIC OCEAN
ATLANTIC OCEAN
SOUTHERN OCEAN
ARCTIC OCEAN
ASIA
GREENLAND
HUDSON BAY
BERING SEA
Gulf of Alaska
ROCKY MOUNTAINS
GREAT PLAINS
CENTRAL LOWLAND
GULF OF MEXICO
CARIBBEAN SEA
WEST INDIES
BAHAMA ISLANDS
BERMUDA
HAWAIIAN ISLANDS
MIDWAY IS.
ALEUTIAN ISLANDS
NEWFOUNDLAND
C. MENDOCINO
VANCOUVER I.
PENINSULA DE BAJA CALIFORNIA
C. SAN LUCAS
C. HATTERAS
C. COD
C. SABLE
REVILLAGIGEDO
Clipperton
ARCH. DE COLÓN (GALÁPAGOS IS.)
SELVAS
CAMPOS
PAMPAS
PATAGONIA
FALKLAND IS.
TIERRA DEL FUEGO
CABO DE HORNOS
SOUTH GEORGIA
SOUTH SANDWICH IS.
SOUTH ORKNEY IS.
SOUTH SHETLAND IS.
WEDDELL SEA
ROSS SEA
Marie Byrd Land
South Pole
North Pole
MARQUESAS IS.
SOCIETY IS.
COOK IS.
TUAMOTU
SAMOA
TONGA
FIJI
KERMADEC IS.
CHATHAM IS.
PHOENIX ISLANDS
TOKELAU IS.
Johnston
Howland
Baker
Jarvis
Malden
Starbuck
Pitcairn
Ducie
Rapa
I. Sala y Gómez
Isla de Pascua (Easter)
I. San Félix
IS. DE JUAN FERNANDEZ
ARCH. DE LOS CHONOS
AZORES (AÇORES)
MADEIRA
IS. CANARIAS
ARQUIPÉLAGO DE CABO VERDE
C. VERT
C. PALMAS
ASCENSION
ST. HELENA
TRISTAN DA CUNHA
GOUGH
KAP FARVEL
ICELAND
Tropic of Cancer
Equator
Longitude West of Greenwich
Scale 1 : 100 000 000 (approximate)
One inch to 1,600 miles
0 500 1000 1500 2000 miles
0 500 1000 1500 2000 2500 Kilometers
Meters Feet
3 050 10 000
1 525 5 000
610 2 000
305 1 000
0 SEA L. 0
BELOW SEA LEVEL
152.5 500
3 050 10 000
6 100 20 000
Land Elevations in Profile
OCEANIA
NORTH AMERICA
SOUTH AMERICA
AFRICA
NEW ZEALAND
HAWAI'I
ALASKA RANGE
CASCADE RANGE
SIERRA NEVADA
ROCKY MTS.
LOS ANDES
ATLAS
PLATEAU OF BOLIVIA
Feet Meters
Ocean Depths in Profile
PACIFIC OCEAN
ATLANTIC OCEA
INDOCHINA
HAINAN
MARIANA IS.
HAWAII
MEXICO
NOVA SCOTIA
Sea Level
A Section along 20° N. Lat.
A Section along 45° N. Lat.
Feet Meters
Elevations and depressions

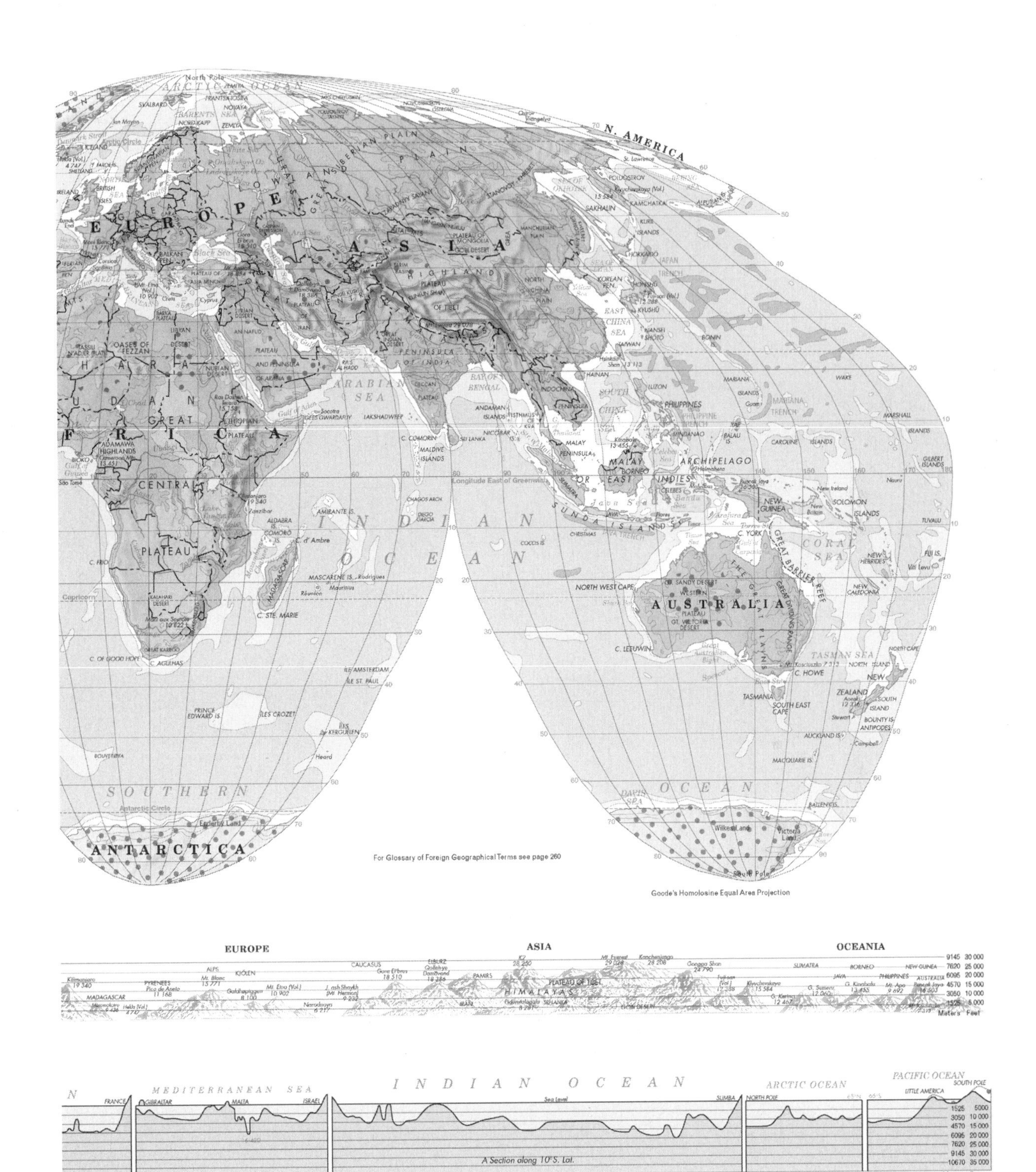


are given in feet

Guilin, China *Corbis Digital Stock*

# World Physical Map

Name: Ryan Ratkovsky Date: ______

See pages 4-5 in Goode's World Atlas (21e) for full color maps.

What are the major **mountain ranges** on each continent?

North America Rocky Mountains, Appalachians, Cascade, Coast Mountains
South America Andes, Brazilian Highlands
Europe Appennino, Carpathians
Asia
Africa
Australia Great Dividing Range, Darling Range, King Leopold Ranges, Hamersely Range

What are the major **rivers** on each continent?

North America Missippi, Yukon, Mackenzie, St. Lawrence, Missouri, Saskatchewan
South America Amazon, Parana
Europe
Asia Yangtze, Huang, Lena, Yenisey
Africa Nile, Congo
Australia Darling

List the **continent(s)** where you find each of the following features.

Hudson Bay North America
Istmo (Isthmus) of Tehuantepec North America
Llanos South America
Cabo de Hornos (Cape Horn) South America
Greater Antilles (islands) North America
Cape of Good Hope ~~Africa~~ Africa
Lake Victoria Africa
Lake Chad Africa
Madagascar Africa
Plateau of Tibet Asia
Great Siberian Plain Asia
Black Sea Europe
Caspian Sea Europe
Great Sandy Desert Australia
Gulf of Oman Africa
Aral Sea Asia
North Sea Europe

Look at the latitudes (north or south of the equator) of the world's major deserts. Is there a pattern? If so, what might explain it?

Yes, these areas weren't carved w/ glaciers

What are the major island groups of the Pacific Ocean?

East Indies

What are the major island groups of the Atlantic Ocean?

West Indies

What are the major island groups of the Indian Ocean?

Toledo, Spain *Corbis Digital Stock*

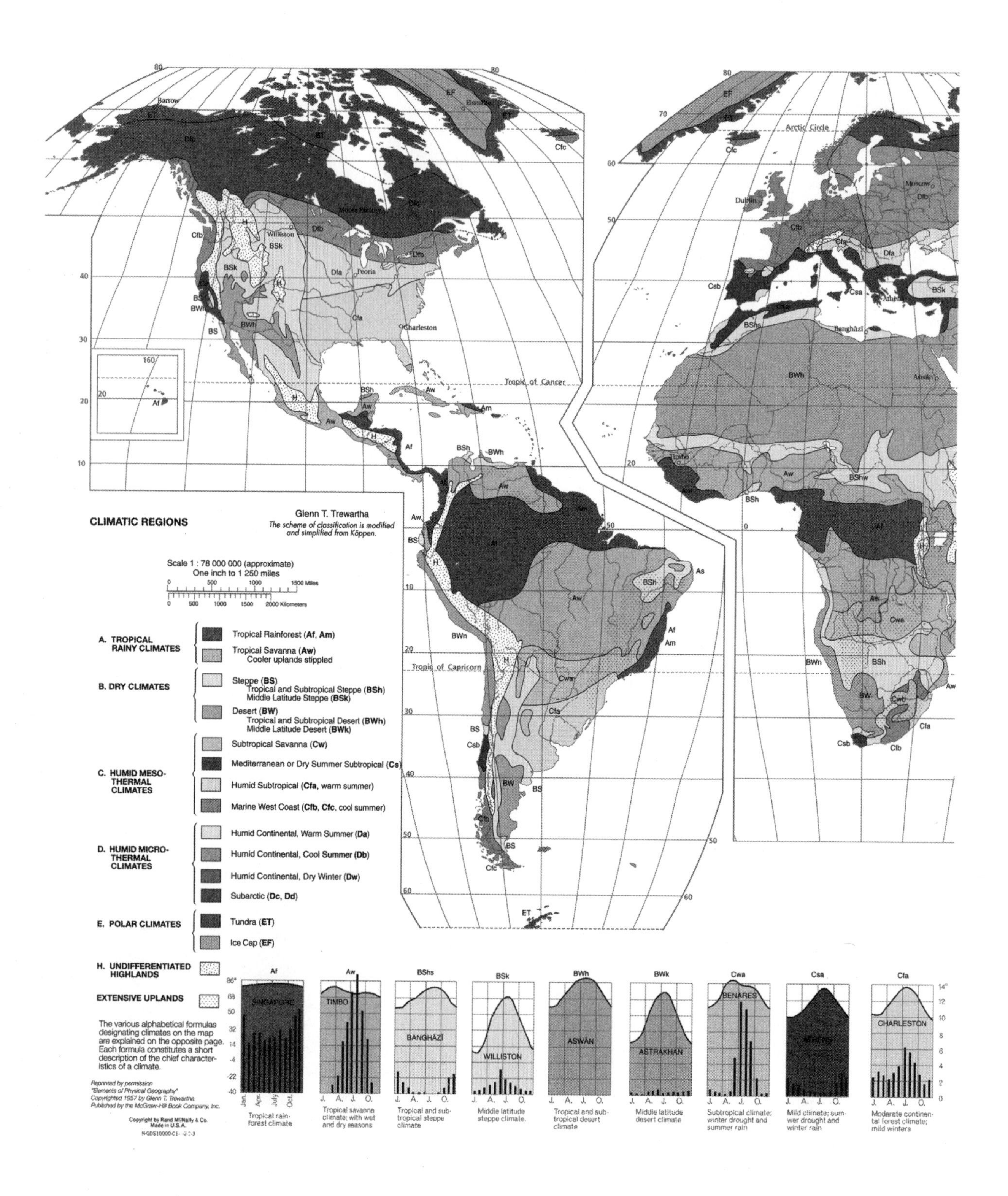

CLIMATIC REGIONS
Glenn T. Trewartha
The scheme of classification is modified and simplified from Köppen.
Scale 1 : 78 000 000 (approximate)
One inch to 1 250 miles
0 500 1000 1500 Miles
0 500 1000 1500 2000 Kilometers
A. TROPICAL RAINY CLIMATES
Tropical Rainforest (Af, Am)
Tropical Savanna (Aw)
Cooler uplands stippled
B. DRY CLIMATES
Steppe (BS)
Tropical and Subtropical Steppe (BSh)
Middle Latitude Steppe (BSk)
Desert (BW)
Tropical and Subtropical Desert (BWh)
Middle Latitude Desert (BWk)
C. HUMID MESO-THERMAL CLIMATES
Subtropical Savanna (Cw)
Mediterranean or Dry Summer Subtropical (Cs)
Humid Subtropical (Cfa, warm summer)
Marine West Coast (Cfb, Cfc, cool summer)
D. HUMID MICRO-THERMAL CLIMATES
Humid Continental, Warm Summer (Da)
Humid Continental, Cool Summer (Db)
Humid Continental, Dry Winter (Dw)
Subarctic (Dc, Dd)
E. POLAR CLIMATES
Tundra (ET)
Ice Cap (EF)
H. UNDIFFERENTIATED HIGHLANDS
EXTENSIVE UPLANDS
The various alphabetical formulas designating climates on the map are explained on the opposite page. Each formula constitutes a short description of the chief characteristics of a climate.
Reprinted by permission
"Elements of Physical Geography"
Copyrighted 1957 by Glenn T. Trewartha.
Published by the McGraw-Hill Book Company, Inc.
Copyright by Rand McNally & Co.
Made in U.S.A.
Barrow
Williston
Peoria
Charleston
Dublin
Moscow
Athens
Banghazi
Aswān
Timbo
Arctic Circle
Tropic of Cancer
Tropic of Capricorn
Af
SINGAPORE
Tropical rain-forest climate
Aw
TIMBO
Tropical savanna climate; with wet and dry seasons
BShs
BANGHĀZĪ
Tropical and sub-tropical steppe climate
BSk
WILLISTON
Middle latitude steppe climate.
BWh
ASWĀN
Tropical and sub-tropical desert climate
BWk
ASTRAKHAN
Middle latitude desert climate
Cwa
BENARES
Subtropical climate; winter drought and summer rain
Csa
ATHENS
Mild climate; summer drought and winter rain
Cfa
CHARLESTON
Moderate continental forest climate; mild winters
86° 68 50 32 14 -4 -22 -40
14" 12 10 8 6 4 2 0
Jan. Apr. July Oct.
J. A. J. O.

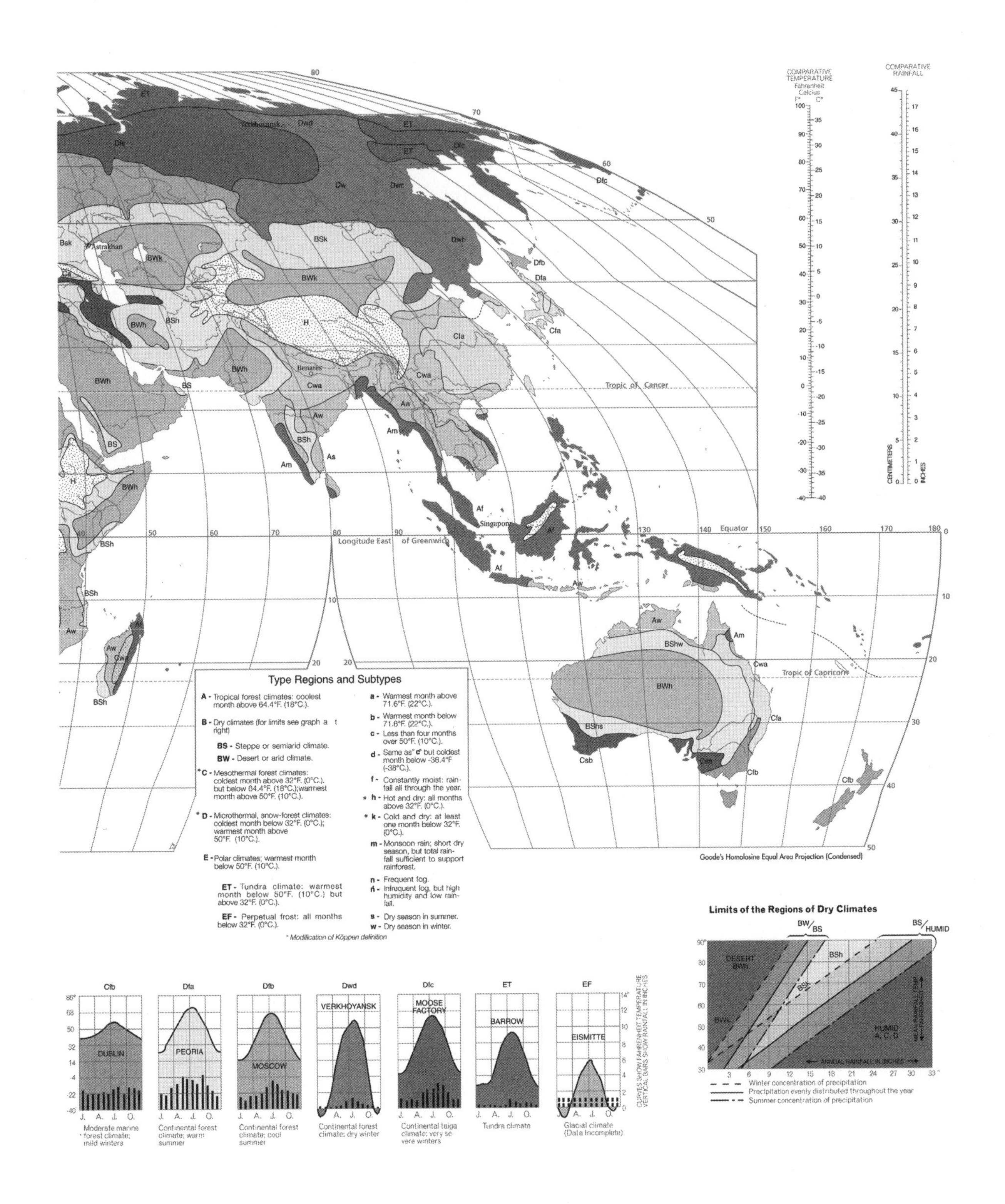
COMPARATIVE TEMPERATURE
Fahrenheit Celcius
COMPARATIVE RAINFALL
CENTIMETERS
INCHES
Verkhoyansk
Astrakhan
Benares
Singapore
Tropic of Cancer
Equator
Tropic of Capricorn
Longitude East of Greenwich
Goode's Homolosine Equal Area Projection (Condensed)
Type Regions and Subtypes
A - Tropical forest climates: coolest month above 64.4°F. (18°C.).
B - Dry climates (for limits see graph at right)
BS - Steppe or semiarid climate.
BW - Desert or arid climate.
*C - Mesothermal forest climates: coldest month above 32°F. (0°C.). but below 64.4°F. (18°C.);warmest month above 50°F. (10°C.).
* D - Microthermal, snow-forest climates: coldest month below 32°F. (0°C.); warmest month above 50°F. (10°C.).
E - Polar climates; warmest month below 50°F. (10°C.).
ET - Tundra climate: warmest month below 50°F. (10°C.) but above 32°F. (0°C.).
EF - Perpetual frost: all months below 32°F. (0°C.).
a - Warmest month above 71.6°F. (22°C.).
b - Warmest month below 71.6°F. (22°C.).
c - Less than four months over 50°F. (10°C.).
d - Same as "c" but coldest month below -36.4°F (-38°C.).
f - Constantly moist: rainfall all through the year.
* h - Hot and dry: all months above 32°F. (0°C.).
* k - Cold and dry: at least one month below 32°F. (0°C.).
m - Monsoon rain; short dry season, but total rainfall sufficient to support rainforest.
n - Frequent fog.
ń - Infrequent fog, but high humidity and low rainfall.
s - Dry season in summer.
w - Dry season in winter.
* Modification of Köppen definition
Cfb
DUBLIN
Moderate marine forest climate; mild winters
Dfa
PEORIA
Continental forest climate; warm summer
Dfb
MOSCOW
Continental forest climate; cool summer
Dwd
VERKHOYANSK
Continental forest climate: dry winter
Dfc
MOOSE FACTORY
Continental taiga climate; very severe winters
ET
BARROW
Tundra climate
EF
EISMITTE
Glacial climate (Data Incomplete)
CURVES SHOW FAHRENHEIT TEMPERATURE
VERTICAL BARS SHOW RAINFALL IN INCHES
Limits of the Regions of Dry Climates
BW/BS
BS/HUMID
DESERT BWh
BWk
BSh
BSk
HUMID A, C, D
ANNUAL RAINFALL IN INCHES
MEAN ANNUAL TEMP. FAHRENHEIT
Winter concentration of precipitation
Precipitation evenly distributed throughout the year
Summer concentration of precipitation

Alaska, United States *Corbis Images*

# *World Climate*

Name: ______________________ Date: ______________________

See pages **14-15** in Goode's World Atlas (21e) for full color maps.

In general, where are **Tropical Rainy Climates** (A) located?

Close to the equator

Central america

Northern part of south america

east indies

In general, where are **Dry Climates** (B) located?

on the equator

Northern africa

australia

In general, where are **Humid Mesothermal Climates** (C) located?

europe

In general, where are **Humid Microthermal Climates** (D) located?

Russia

North America

In general, where are **Polar Climates** (E) located?

artic

What climate types are found along the Equator?

tropical rainy climates

Do continents have the same climate types on their eastern and western coasts? What might explain the similarities or differences?

europe-yes

others no- size of the continent
- the environment is different on either coast

What are the links between elevation and the location of H climates?

Death Valley, California, USA *Corbis Digital Stock*

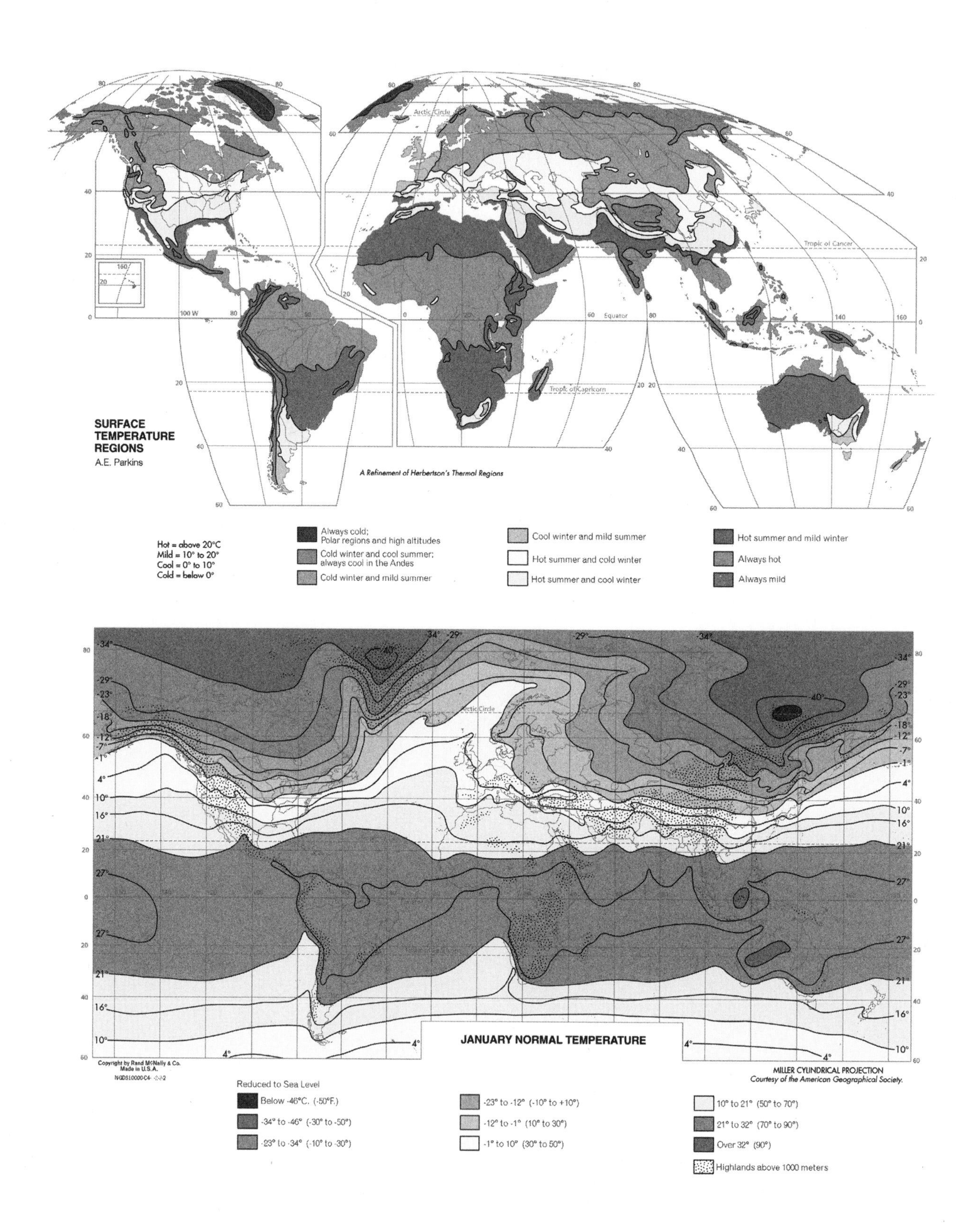
SURFACE TEMPERATURE REGIONS
A.E. Parkins
A Refinement of Herbertson's Thermal Regions
Arctic Circle
Tropic of Cancer
Equator
Tropic of Capricorn
Hot = above 20°C
Mild = 10° to 20°
Cool = 0° to 10°
Cold = below 0°
Always cold; Polar regions and high altitudes
Cold winter and cool summer; always cool in the Andes
Cold winter and mild summer
Cool winter and mild summer
Hot summer and cold winter
Hot summer and cool winter
Hot summer and mild winter
Always hot
Always mild
JANUARY NORMAL TEMPERATURE
Copyright by Rand McNally & Co. Made in U.S.A.
MILLER CYLINDRICAL PROJECTION
Courtesy of the American Geographical Society.
Reduced to Sea Level
Below -46°C. (-50°F.)
-34° to -46° (-30° to -50°)
-23° to -34° (-10° to -30°)
-23° to -12° (-10° to +10°)
-12° to -1° (10° to 30°)
-1° to 10° (30° to 50°)
10° to 21° (50° to 70°)
21° to 32° (70° to 90°)
Over 32° (90°)
Highlands above 1000 meters

## *World Surface Temperature Regions*

Name: ______________________________ Date: ______________________________

See page **16** in Goode's World Atlas (21e) for a full color map.

What is the relationship between latitude and surface temperature?

- at the tropics of cancer & capricorn you find your deserts
- equator find your rainy/humid climates

What connections can you notice between physical geography (elevation, proximity to water, etc.) and temperature patterns?

What factors explain the areas of the world that are "always hot"?

What factors might explain the location of areas that are cold? Is it just a matter of latitude?

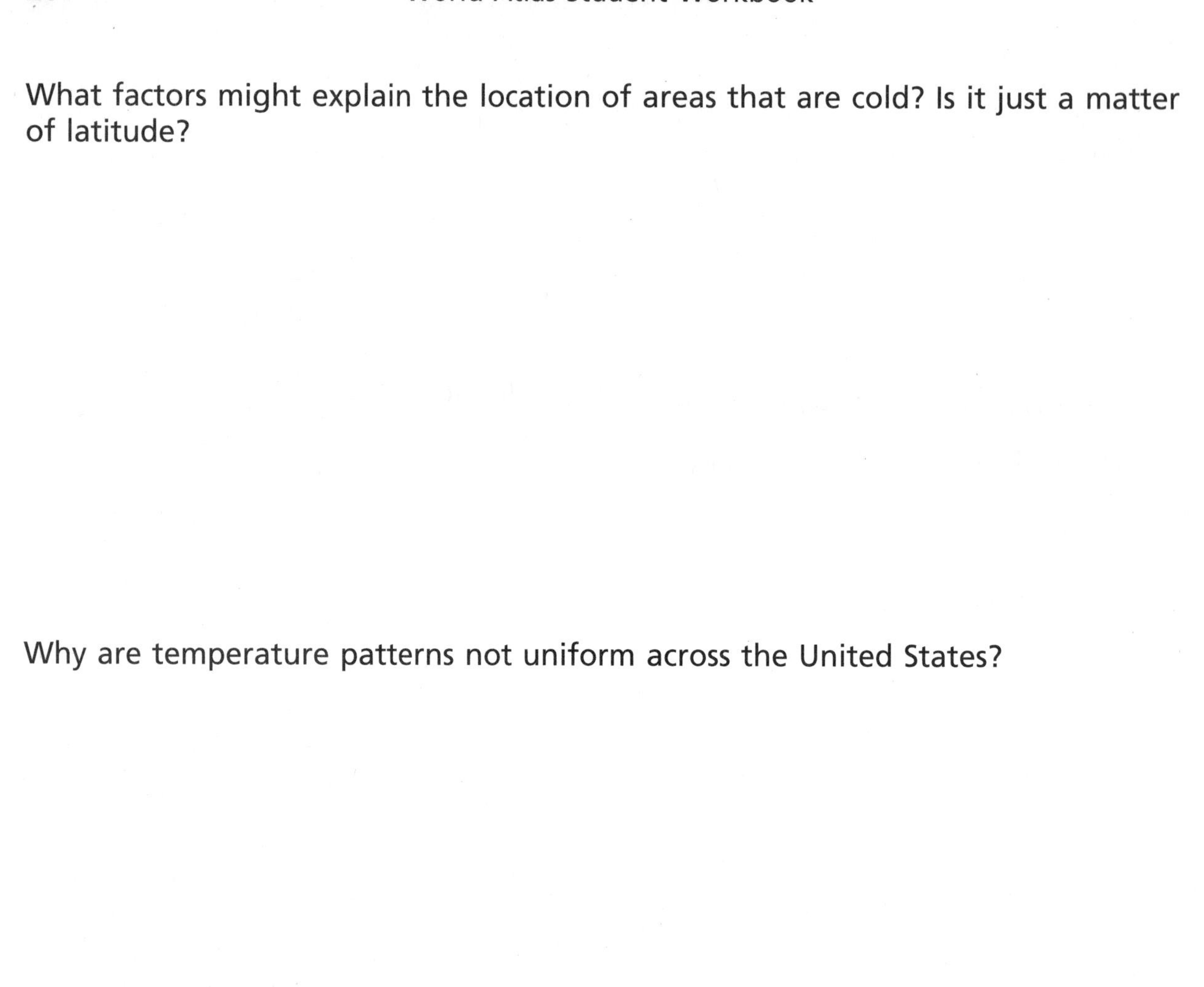

Why are temperature patterns not uniform across the United States?

Coastal areas of northern Europe, such as the coast of Norway, have much milder temperatures than locations of equal latitude in northern Canada. What might explain this?

Java, Indonesia *Corbis Digital Stock*

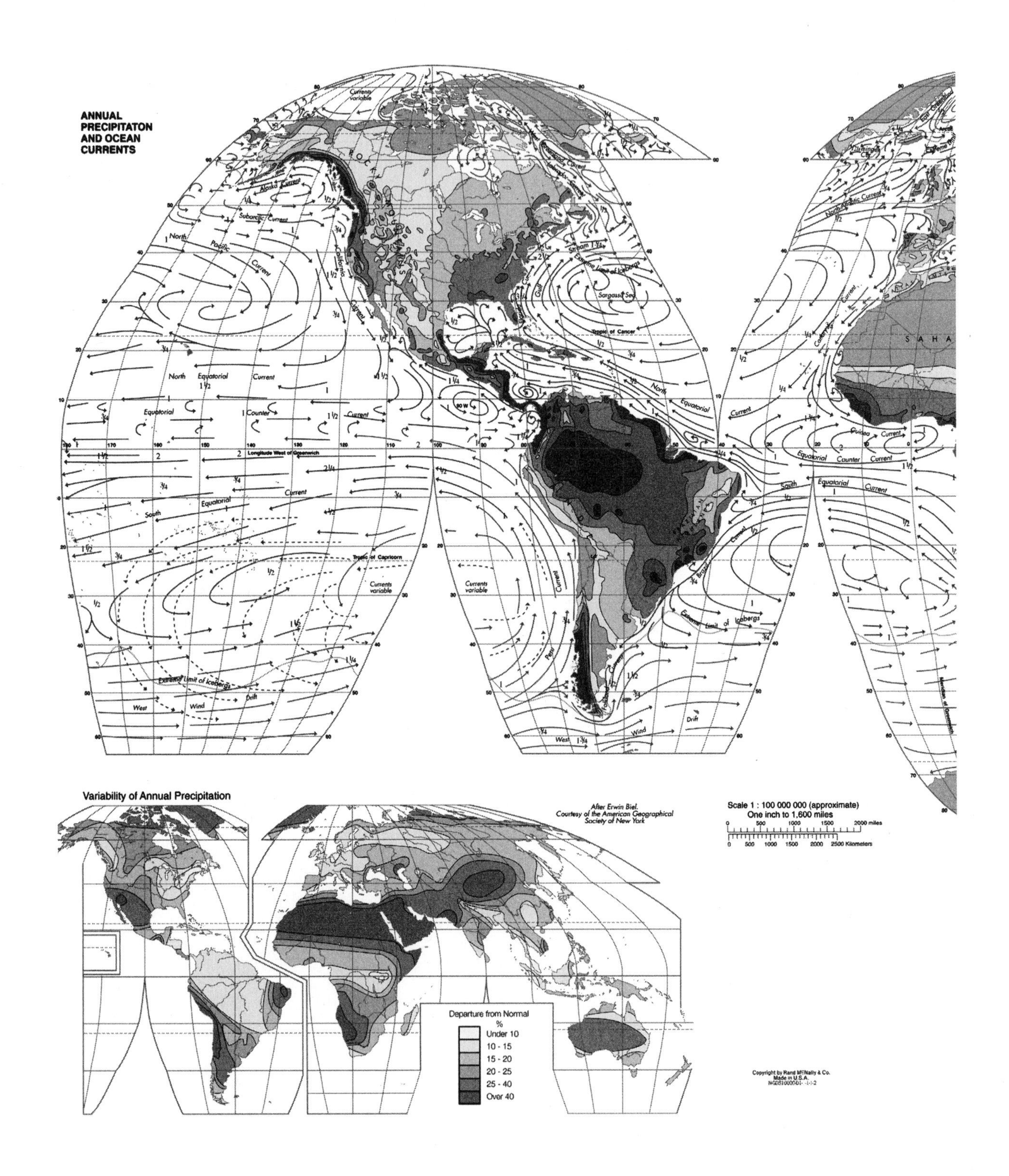
ANNUAL PRECIPITATON AND OCEAN CURRENTS
Variability of Annual Precipitation
After Erwin Biel. Courtesy of the American Geographical Society of New York
Scale 1 : 100 000 000 (approximate)
One inch to 1,600 miles
Departure from Normal %
Under 10
10 - 15
15 - 20
20 - 25
25 - 40
Over 40
North Equatorial Current
Equatorial Counter Current
South Equatorial Current
North Pacific Current
Alaska Current
Subarctic Current
California Current
Gulf Stream
Sargasso Sea
Tropic of Cancer
Tropic of Capricorn
Brazil Current
Peru Current
Extreme Limit of Icebergs
West Wind Drift
Currents variable
Longitude West of Greenwich
Guinea Current
North Atlantic Current
Canaries Current
Copyright by Rand McNally & Co. Made in U.S.A.

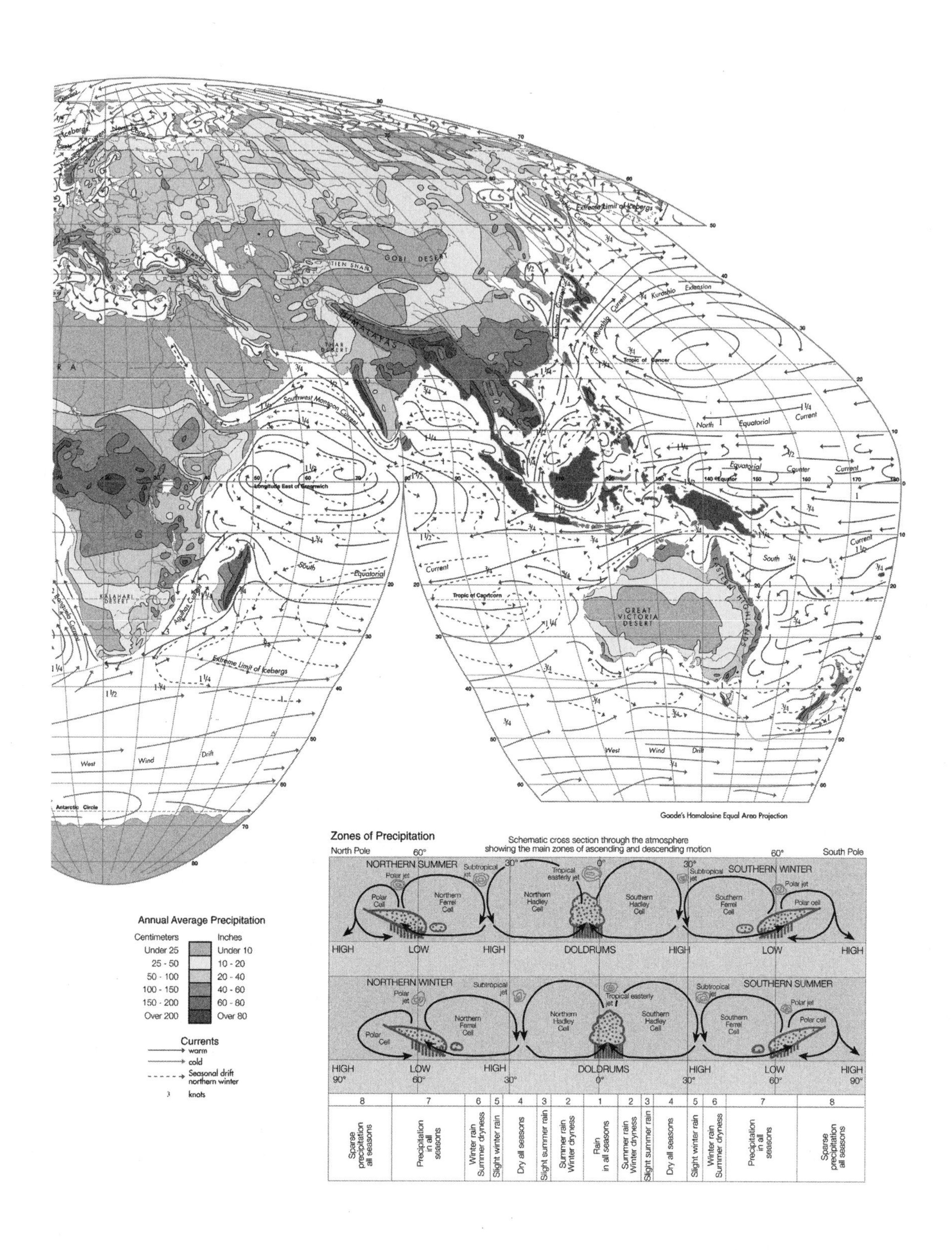
GOBI DESERT
TIEN SHAN
HIMALAYAS
THAR DESERT
KALAHARI DESERT
GREAT VICTORIA DESERT
Extreme Limit of Icebergs
Kuroshio Extension
Tropic of Cancer
North Equatorial Current
Equatorial Counter Current
Southwest Monsoon Current
Longitude East of Greenwich
South Equatorial Current
Tropic of Capricorn
West Wind Drift
Antarctic Circle
Goode's Homolosine Equal Area Projection
Annual Average Precipitation
Centimeters Inches
Under 25 Under 10
25 - 50 10 - 20
50 - 100 20 - 40
100 - 150 40 - 60
150 - 200 60 - 80
Over 200 Over 80
Currents
warm
cold
Seasonal drift northern winter
3 knots
Zones of Precipitation
Schematic cross section through the atmosphere showing the main zones of ascending and descending motion
North Pole 60° 30° 0° 30° 60° South Pole
NORTHERN SUMMER
SOUTHERN WINTER
Polar jet
Subtropical jet
Tropical easterly jet
Polar Cell
Northern Ferrel Cell
Northern Hadley Cell
Southern Hadley Cell
Southern Ferrel Cell
Polar cell
HIGH LOW HIGH DOLDRUMS HIGH LOW HIGH
NORTHERN WINTER
SOUTHERN SUMMER
HIGH 90° LOW 60° HIGH 30° DOLDRUMS 0° HIGH 30° LOW 60° HIGH 90°
8 7 6 5 4 3 2 1 2 3 4 5 6 7 8
Sparse precipitation all seasons
Precipitation in all seasons
Winter rain Summer dryness
Slight winter rain
Dry all seasons
Slight summer rain
Summer rain Winter dryness
Rain in all seasons
Summer rain Winter dryness
Slight summer rain
Dry all seasons
Slight winter rain
Winter rain Summer dryness
Precipitation in all seasons
Sparse precipitation all seasons

Arizona, United States *Corbis Digital Stock*

## *World Precipitation*

Name: ______________________________ Date: ____________________________

See pages **20-21** in Goode's World Atlas (21e) for full color maps.

---

What areas of the world receive the most precipitation?

What areas of the world receive the least precipitation?

What might explain the low levels of precipitation along the western coasts of South America, southern Africa, and western Australia? What do they have in common?

Moving inland from the west coast of the United States, you pass quickly from a very wet area to a very dry area. What factors might explain this change?

Summarize the direction of ocean currents in the Northern and Southern Hemispheres. Are the patterns the same in each hemisphere?

Can you see find any relationship between warm ocean currents and precipitation? Where?

Most of Indonesia receives a lot of rain, but similar latitudes in Africa are drier. What might explain this difference?

*Corbis Digital Stock*

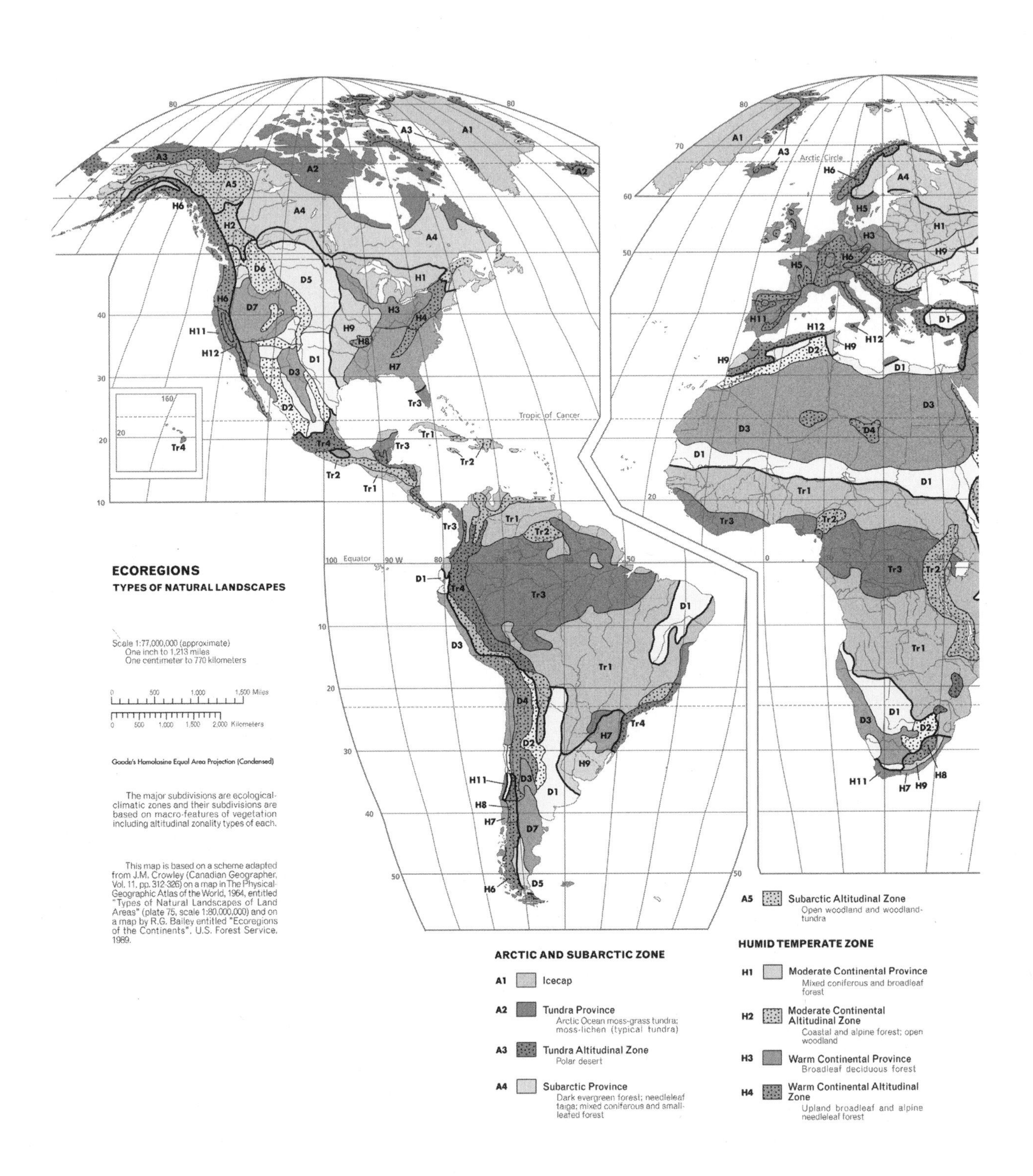
ECOREGIONS
TYPES OF NATURAL LANDSCAPES
Scale 1:77,000,000 (approximate)
One inch to 1,213 miles
One centimeter to 770 kilometers
0 500 1,000 1,500 Miles
0 500 1,000 1,500 2,000 Kilometers
Goode's Homolosine Equal Area Projection (Condensed)
The major subdivisions are ecological-climatic zones and their subdivisions are based on macro-features of vegetation including altitudinal zonality types of each.
This map is based on a scheme adapted from J.M. Crowley (Canadian Geographer, Vol. 11, pp. 312-326) on a map in The Physical-Geographic Atlas of the World, 1964, entitled "Types of Natural Landscapes of Land Areas" (plate 75, scale 1:80,000,000) and on a map by R.G. Bailey entitled "Ecoregions of the Continents", U.S. Forest Service, 1989.
Arctic Circle
Tropic of Cancer
Equator
ARCTIC AND SUBARCTIC ZONE
A1 Icecap
A2 Tundra Province
Arctic Ocean moss-grass tundra; moss-lichen (typical tundra)
A3 Tundra Altitudinal Zone
Polar desert
A4 Subarctic Province
Dark evergreen forest; needleleaf taiga; mixed coniferous and small-leafed forest
A5 Subarctic Altitudinal Zone
Open woodland and woodland-tundra
HUMID TEMPERATE ZONE
H1 Moderate Continental Province
Mixed coniferous and broadleaf forest
H2 Moderate Continental Altitudinal Zone
Coastal and alpine forest; open woodland
H3 Warm Continental Province
Broadleaf deciduous forest
H4 Warm Continental Altitudinal Zone
Upland broadleaf and alpine needleleaf forest

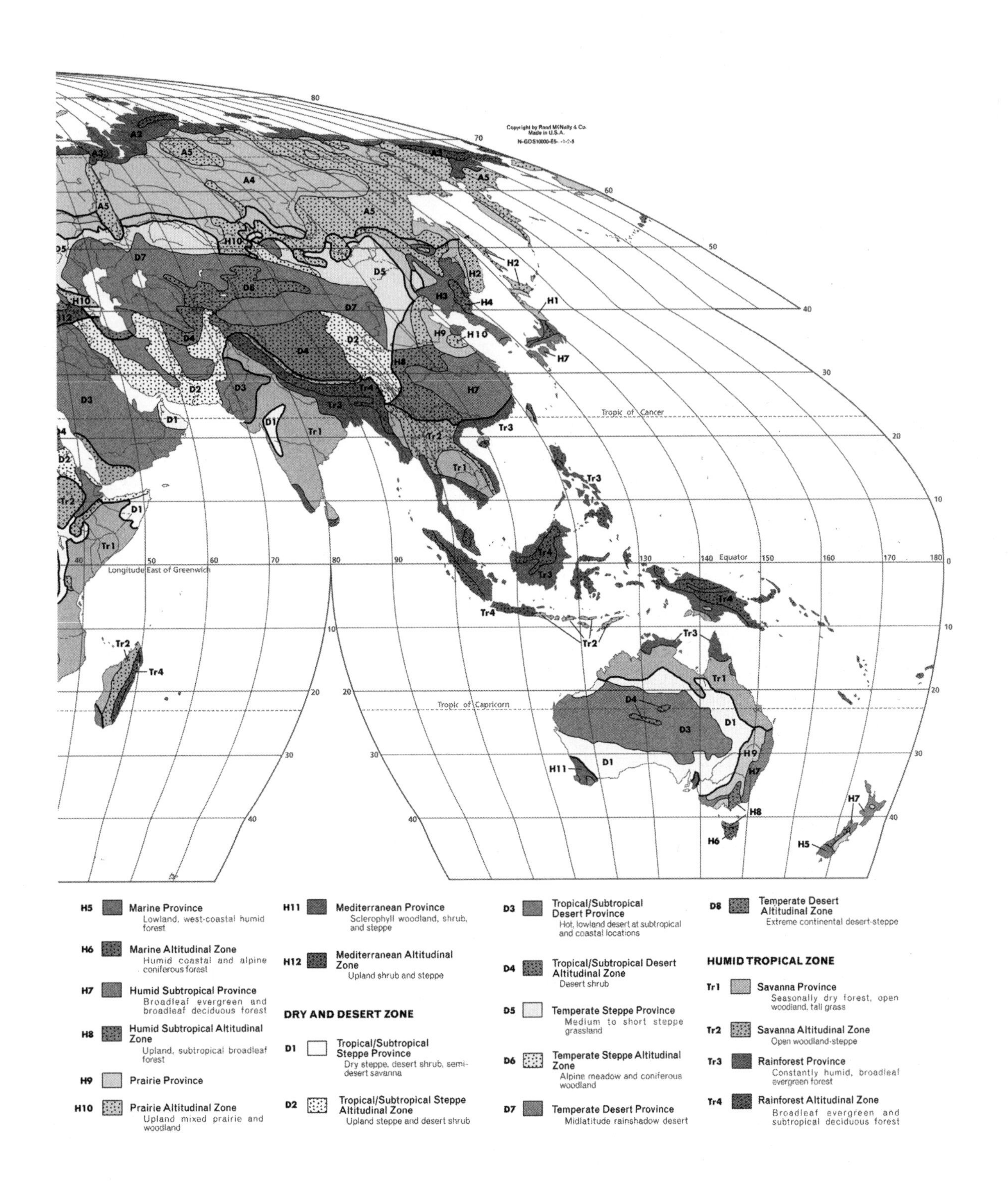
Tropic of Cancer
Equator
Tropic of Capricorn
Longitude East of Greenwich
H5 Marine Province
Lowland, west-coastal humid forest
H6 Marine Altitudinal Zone
Humid coastal and alpine coniferous forest
H7 Humid Subtropical Province
Broadleaf evergreen and broadleaf deciduous forest
H8 Humid Subtropical Altitudinal Zone
Upland, subtropical broadleaf forest
H9 Prairie Province
H10 Prairie Altitudinal Zone
Upland mixed prairie and woodland
H11 Mediterranean Province
Sclerophyll woodland, shrub, and steppe
H12 Mediterranean Altitudinal Zone
Upland shrub and steppe
DRY AND DESERT ZONE
D1 Tropical/Subtropical Steppe Province
Dry steppe, desert shrub, semi-desert savanna
D2 Tropical/Subtropical Steppe Altitudinal Zone
Upland steppe and desert shrub
D3 Tropical/Subtropical Desert Province
Hot, lowland desert at subtropical and coastal locations
D4 Tropical/Subtropical Desert Altitudinal Zone
Desert shrub
D5 Temperate Steppe Province
Medium to short steppe grassland
D6 Temperate Steppe Altitudinal Zone
Alpine meadow and coniferous woodland
D7 Temperate Desert Province
Midlatitude rainshadow desert
D8 Temperate Desert Altitudinal Zone
Extreme continental desert-steppe
HUMID TROPICAL ZONE
Tr1 Savanna Province
Seasonally dry forest, open woodland, tall grass
Tr2 Savanna Altitudinal Zone
Open woodland-steppe
Tr3 Rainforest Province
Constantly humid, broadleaf evergreen forest
Tr4 Rainforest Altitudinal Zone
Broadleaf evergreen and subtropical deciduous forest

Washington State, United States *Corbis Digital Stock*

## *World Ecoregions*

Name: ______________________ Date: ______________________

See pages **28-29** in Goode's World Atlas (21e) for full color maps.

Where in the world are **Mediterranean Provinces** located (H11). What do these areas have in common?

Are there similarities in the location of dry climates worldwide? Explain.

Are there similarities in the location of **Maritime Provinces** (H5) worldwide? Explain.

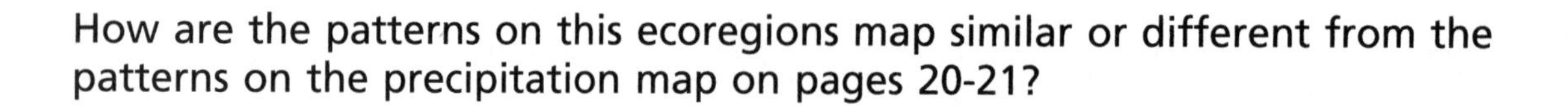

How are the patterns on this ecoregions map similar or different from the patterns on the precipitation map on pages 20-21?

What ecoregions are found along the Equator?

What ecoregions are found north of 60° North latitude?

Hong Kong, China *Corbis Digital Stock*

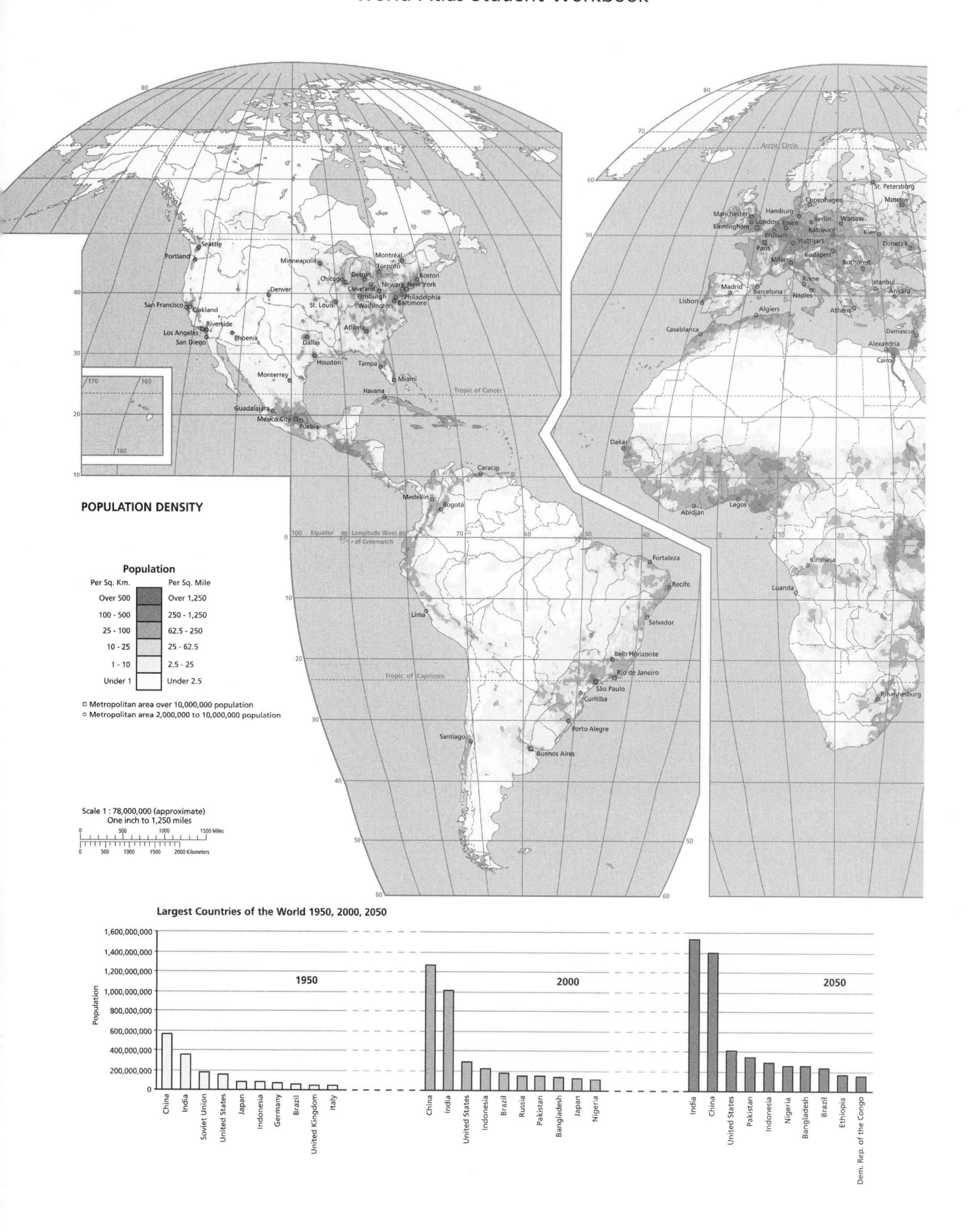


## Largest Countries of the World 1950, 2000, 2050

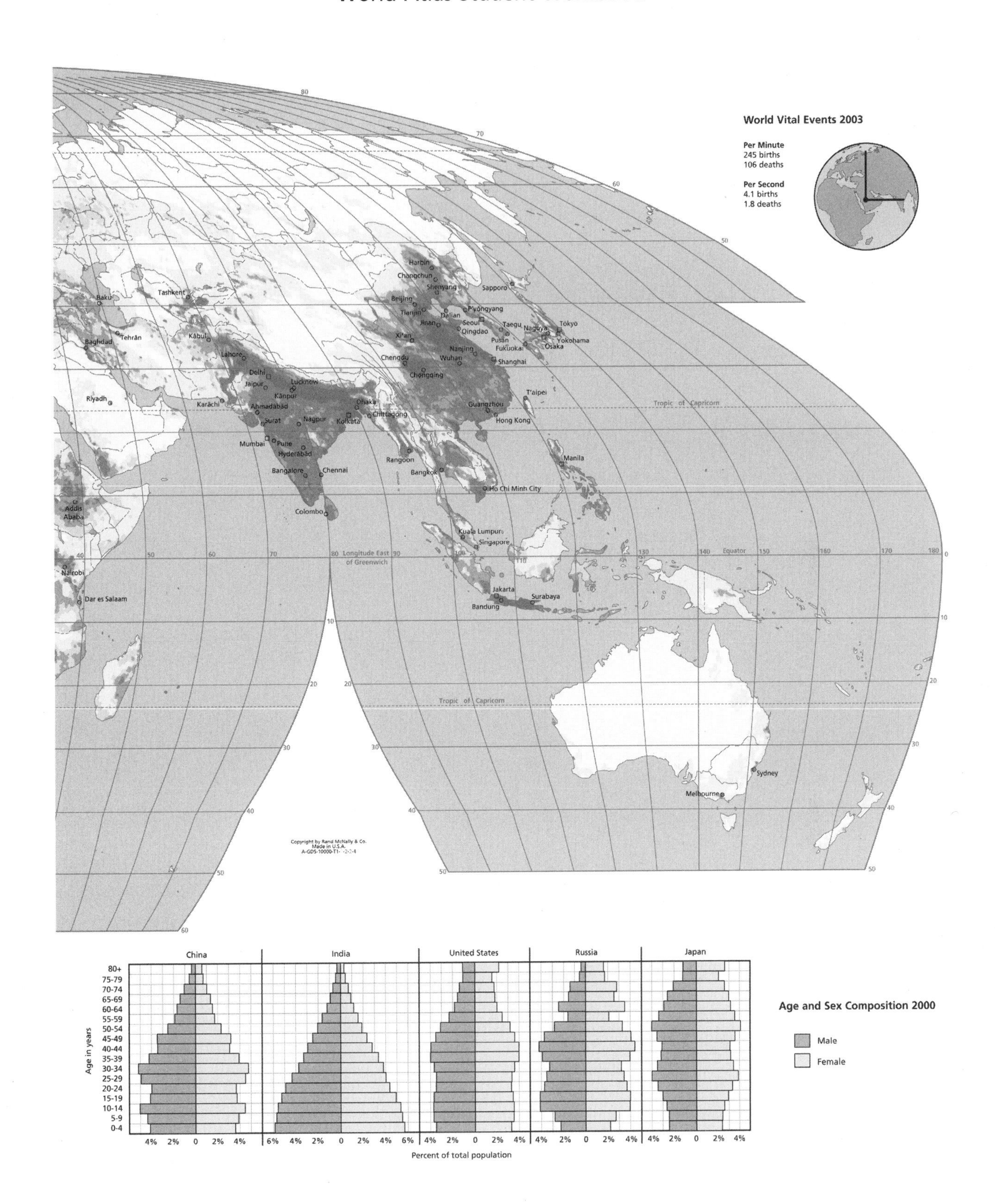
World Vital Events 2003
Per Minute
245 births
106 deaths
Per Second
4.1 births
1.8 deaths
Harbin
Changchun
Shenyang
Sapporo
Beijing
Tianjin
Dalian
P'yongyang
Seoul
Jinan
Qingdao
Taegu
Nagoya
Tokyo
Yokohama
Osaka
Pusan
Fukuoka
Xi'an
Nanjing
Wuhan
Shanghai
Chengdu
Chongqing
T'aipei
Guangzhou
Hong Kong
Baku
Tashkent
Tehrān
Baghdad
Kābul
Lahore
Delhi
Jaipur
Lucknow
Kānpur
Riyadh
Karāchi
Ahmadābād
Surat
Nagpur
Kolkata
Dhaka
Chittagong
Mumbai
Pune
Hyderābād
Bangalore
Chennai
Colombo
Rangoon
Bangkok
Manila
Ho Chi Minh City
Kuala Lumpur
Singapore
Jakarta
Bandung
Surabaya
Addis Ababa
Nairobi
Dar es Salaam
Sydney
Melbourne
Tropic of Capricorn
Equator
Longitude East of Greenwich
China
India
United States
Russia
Japan
Age in years
80+
75-79
70-74
65-69
60-64
55-59
50-54
45-49
40-44
35-39
30-34
25-29
20-24
15-19
10-14
5-9
0-4
Percent of total population
Age and Sex Composition 2000
Male
Female

India *Digital Vision*

## *World Population*

Name: ______________________ Date: ______________________

See pages **30-31** in Goode's World Atlas (21e) for full color maps.

Where are the most densely populated areas of the world?

In the United States, why are more people located in the east rather than the west?

Why do you think there are very few population clusters along the Equator?

What explains the drastic decline in population density as you move north from India into China?

Why is there a large gap in population between Lagos and Algiers in West Africa?

What, if any, connection can you make between elevation and population density? Does the pattern of population in South America follow the same pattern as other continents in terms of elevation?

Mexico *PhotoDisc, Inc./Getty Images*

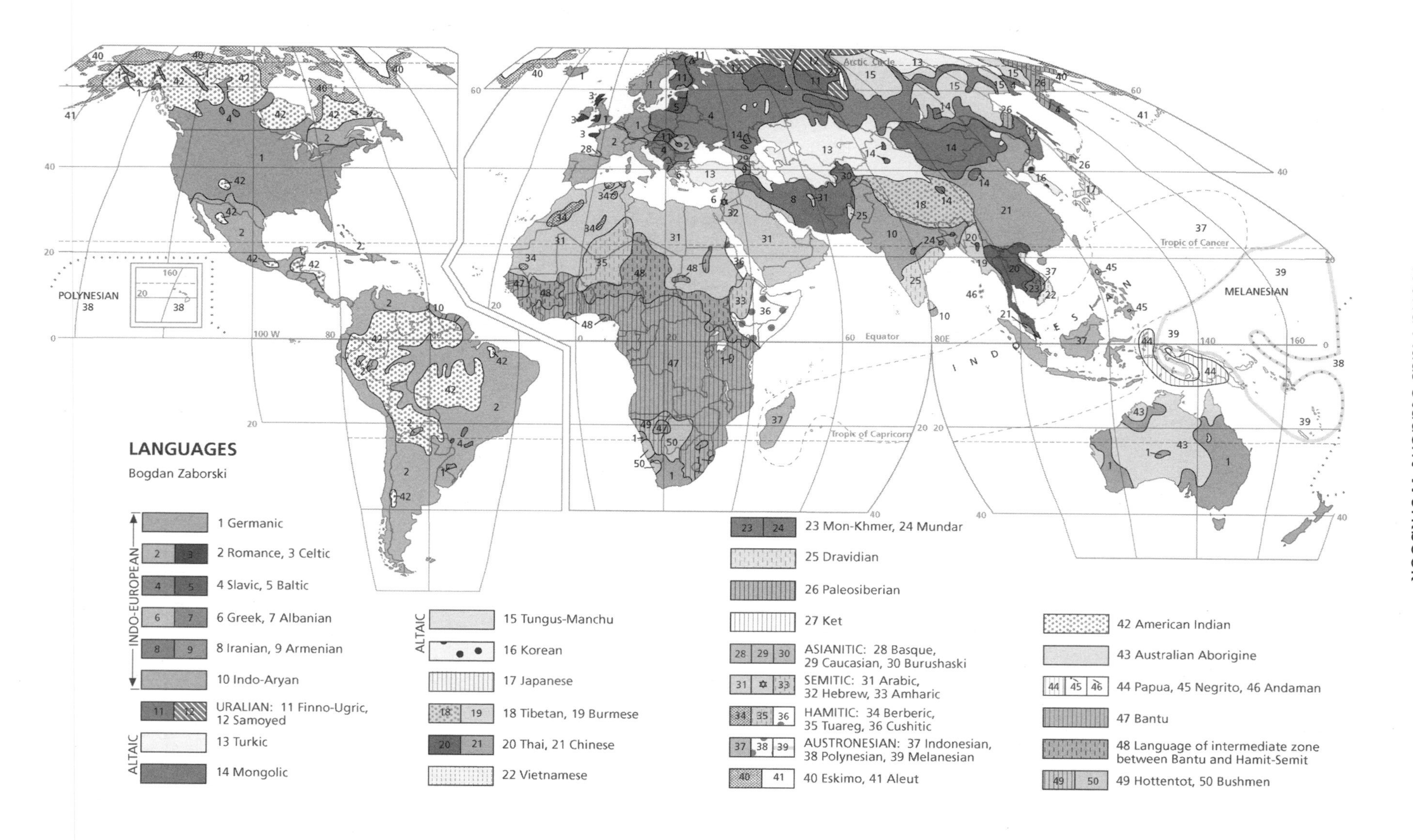
LANGUAGES
Bogdan Zaborski
INDO-EUROPEAN
1 Germanic
2 Romance, 3 Celtic
4 Slavic, 5 Baltic
6 Greek, 7 Albanian
8 Iranian, 9 Armenian
10 Indo-Aryan
URALIAN: 11 Finno-Ugric, 12 Samoyed
ALTAIC
13 Turkic
14 Mongolic
15 Tungus-Manchu
16 Korean
17 Japanese
18 Tibetan, 19 Burmese
20 Thai, 21 Chinese
22 Vietnamese
23 Mon-Khmer, 24 Mundar
25 Dravidian
26 Paleosiberian
27 Ket
ASIANITIC: 28 Basque, 29 Caucasian, 30 Burushaski
SEMITIC: 31 Arabic, 32 Hebrew, 33 Amharic
HAMITIC: 34 Berberic, 35 Tuareg, 36 Cushitic
AUSTRONESIAN: 37 Indonesian, 38 Polynesian, 39 Melanesian
40 Eskimo, 41 Aleut
42 American Indian
43 Australian Aborigine
44 Papua, 45 Negrito, 46 Andaman
47 Bantu
48 Language of intermediate zone between Bantu and Hamit-Semit
49 Hottentot, 50 Bushmen
POLYNESIAN
MELANESIAN
Arctic Circle
Tropic of Cancer
Equator
Tropic of Capricorn
INDONESIA
100 W
80E

## *World Language*

Name: ______________________________ Date: ______________________________

See page **35** in Goode's World Atlas (21e) for a full color map.

---

Why does the Western Hemisphere have so few language families while the Eastern Hemisphere has so many?

Why do you think northern and southern Africa have different major language families?

Where are **Austronesian** languages found? What is unusual about this distribution?

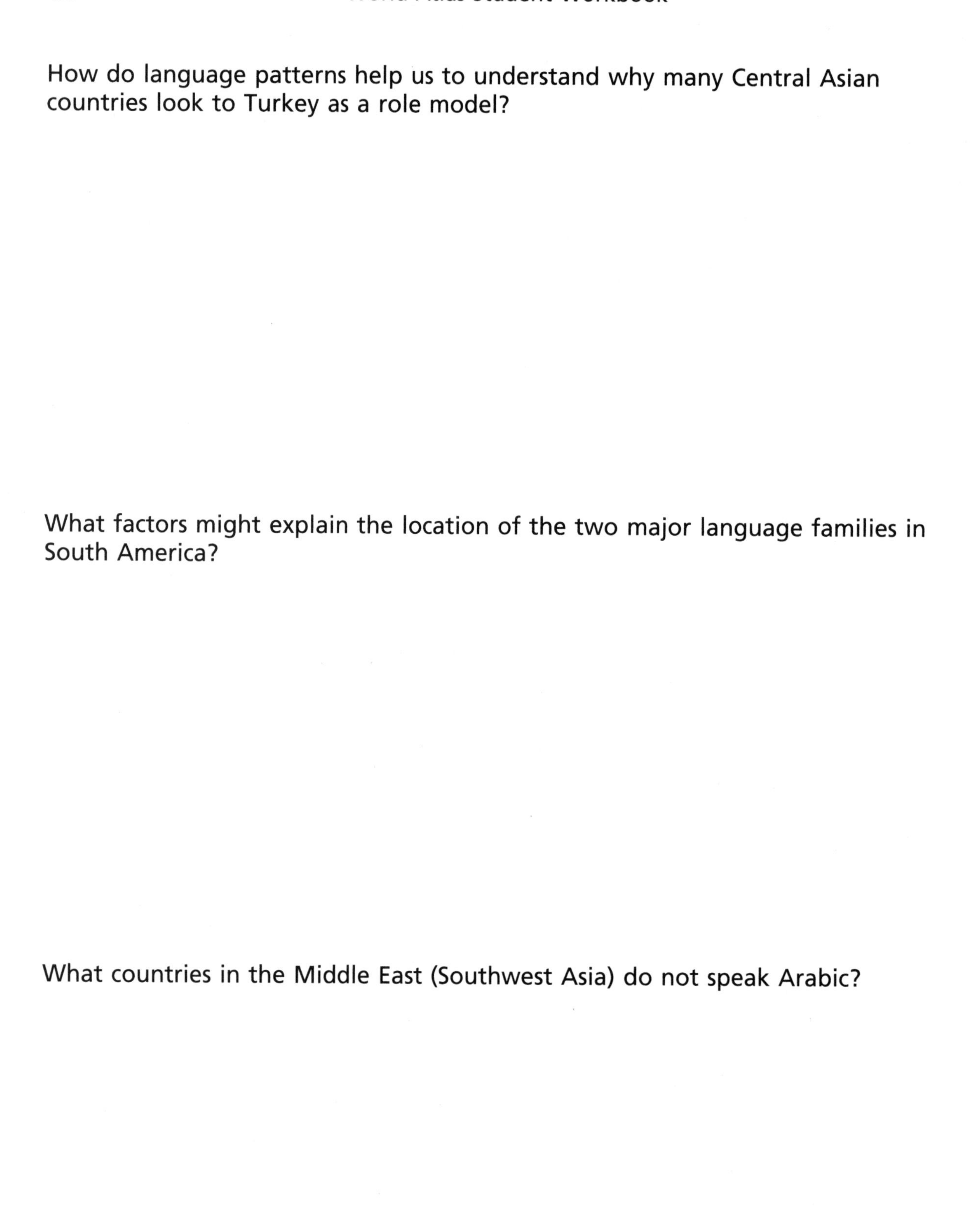

How do language patterns help us to understand why many Central Asian countries look to Turkey as a role model?

What factors might explain the location of the two major language families in South America?

What countries in the Middle East (Southwest Asia) do not speak Arabic?

Kamakura, Japan *Corbis Digital Stock*

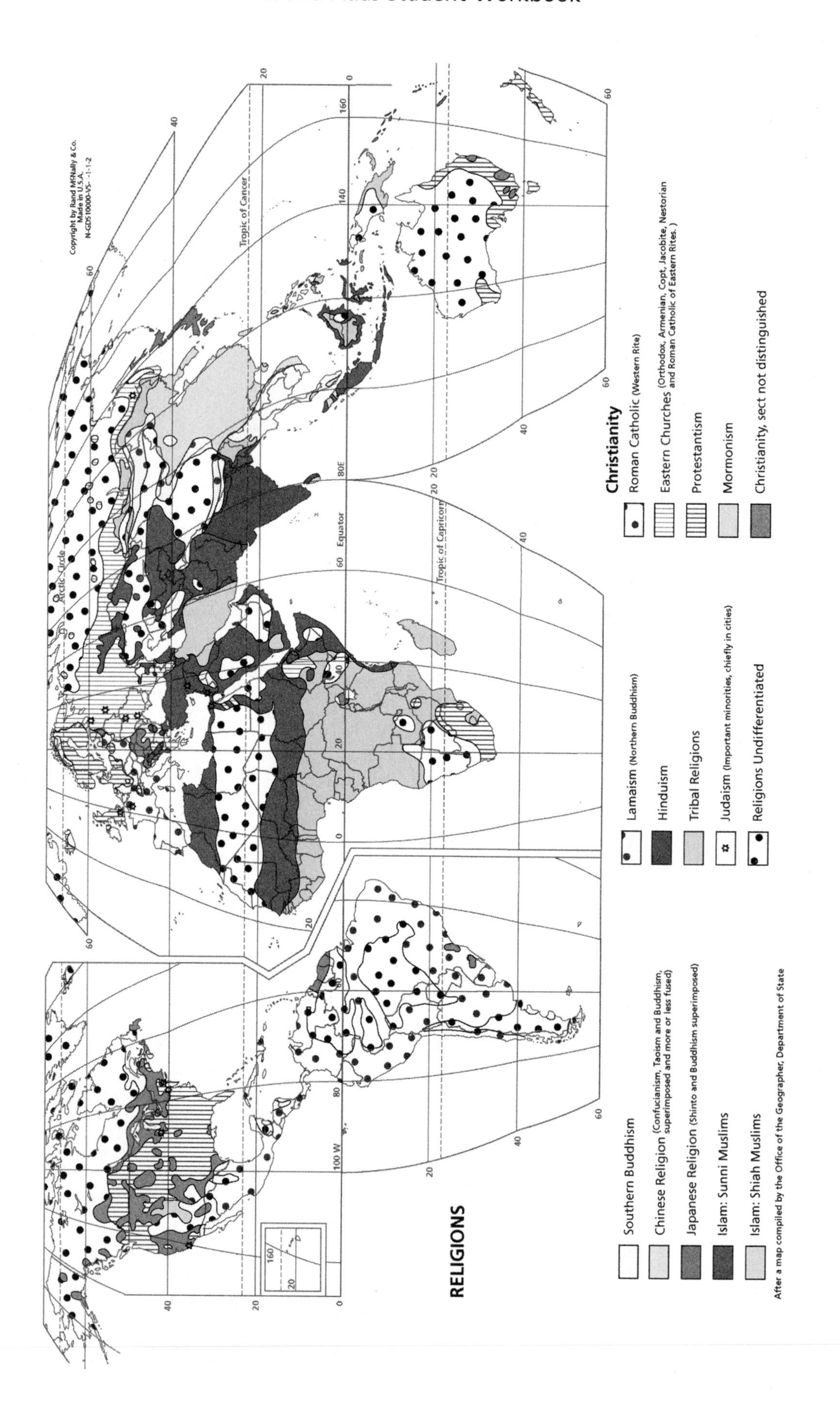
RELIGIONS
Southern Buddhism
Chinese Religion (Confucianism, Taoism and Buddhism, superimposed and more or less fused)
Japanese Religion (Shinto and Buddhism superimposed)
Islam: Sunni Muslims
Islam: Shiah Muslims
Lamaism (Northern Buddhism)
Hinduism
Tribal Religions
Judaism (Important minorities, chiefly in cities)
Religions Undifferentiated
Christianity
Roman Catholic (Western Rite)
Eastern Churches (Orthodox, Armenian, Copt, Jacobite, Nestorian and Roman Catholic of Eastern Rites.)
Protestantism
Mormonism
Christianity, sect not distinguished
Arctic Circle
Tropic of Cancer
Equator
Tropic of Capricorn
100 W
80E
After a map compiled by the Office of the Geographer, Department of State
Copyright by Rand McNally & Co. Made in U.S.A.

## *World Religion*

Name: ______________________ Date: ______________________

See page **35** in Goode's World Atlas (21e) for a full color map.

Describe the distribution of **Islam** around the world.

Describe the distribution of **Christian** areas around the world.

What areas of the world have large territories not dominated by Christianity or Islam?

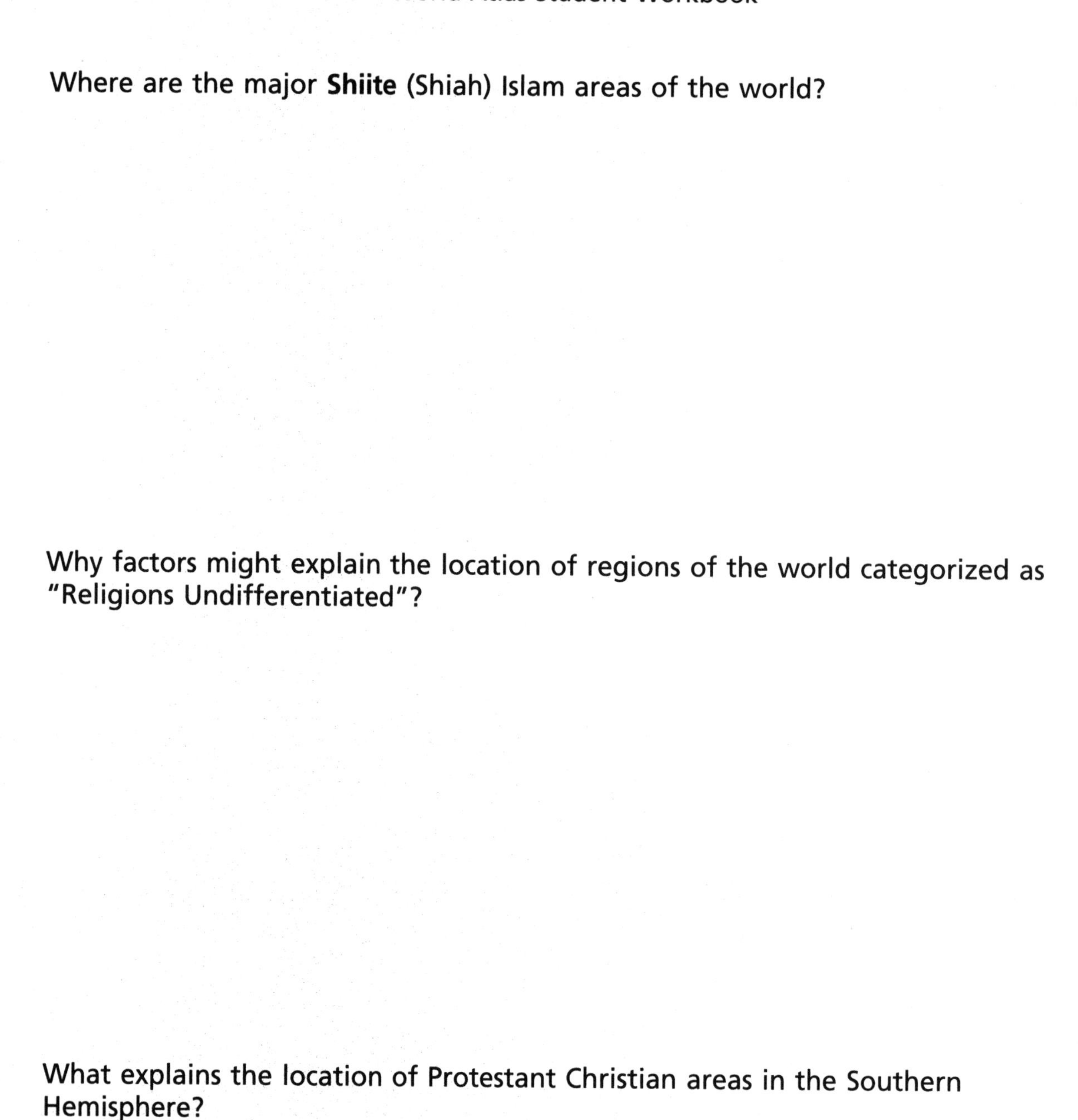

Where are the major **Shiite** (Shiah) Islam areas of the world?

Why factors might explain the location of regions of the world categorized as "Religions Undifferentiated"?

What explains the location of Protestant Christian areas in the Southern Hemisphere?

Quebec City, Canada *Corbis Digital Stock*

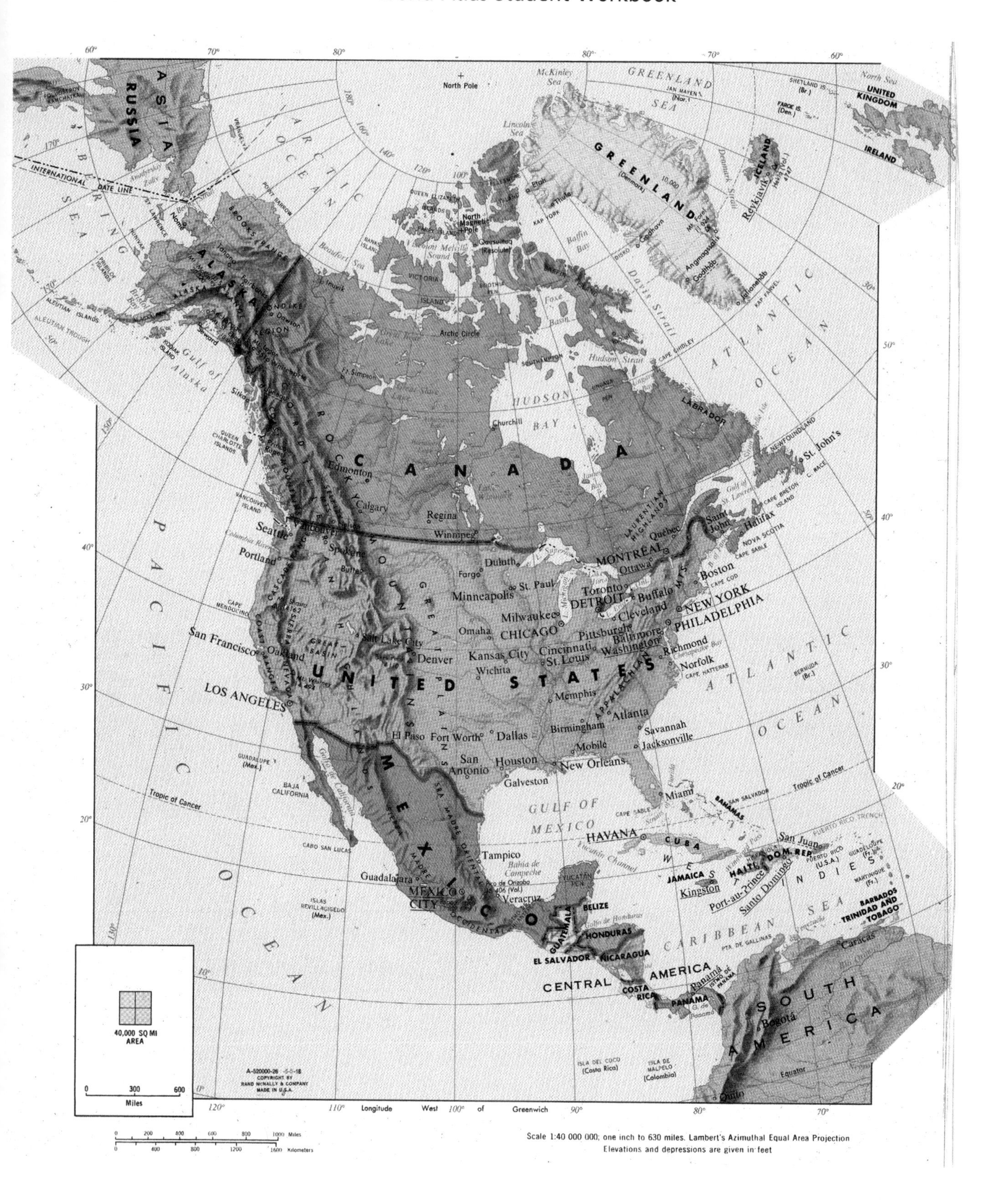


Scale 1:40 000 000; one inch to 630 miles. Lambert's Azimuthal Equal Area Projection
Elevations and depressions are given in feet

## *North America Political Map*

Name: ______________________ Date: ______________________

See page **88** in Goode's World Atlas (21e) for a full color map.

List a **city** for each of the following letters.

| | |
|---|---|
| A ______________ | M ______________ |
| B ______________ | N ______________ |
| C ______________ | O ______________ |
| D ______________ | P ______________ |
| E ______________ | Q ______________ |
| F ______________ | R ______________ |
| G ______________ | S ______________ |
| H ______________ | T ______________ |
| I ______________ | V ______________ |
| J ______________ | W ______________ |
| K ______________ | |

List all of the countries that border the Pacific Ocean.

List all of the countries that have territory south of the Tropic of Cancer.

Which country shares its borders with the most other countries?

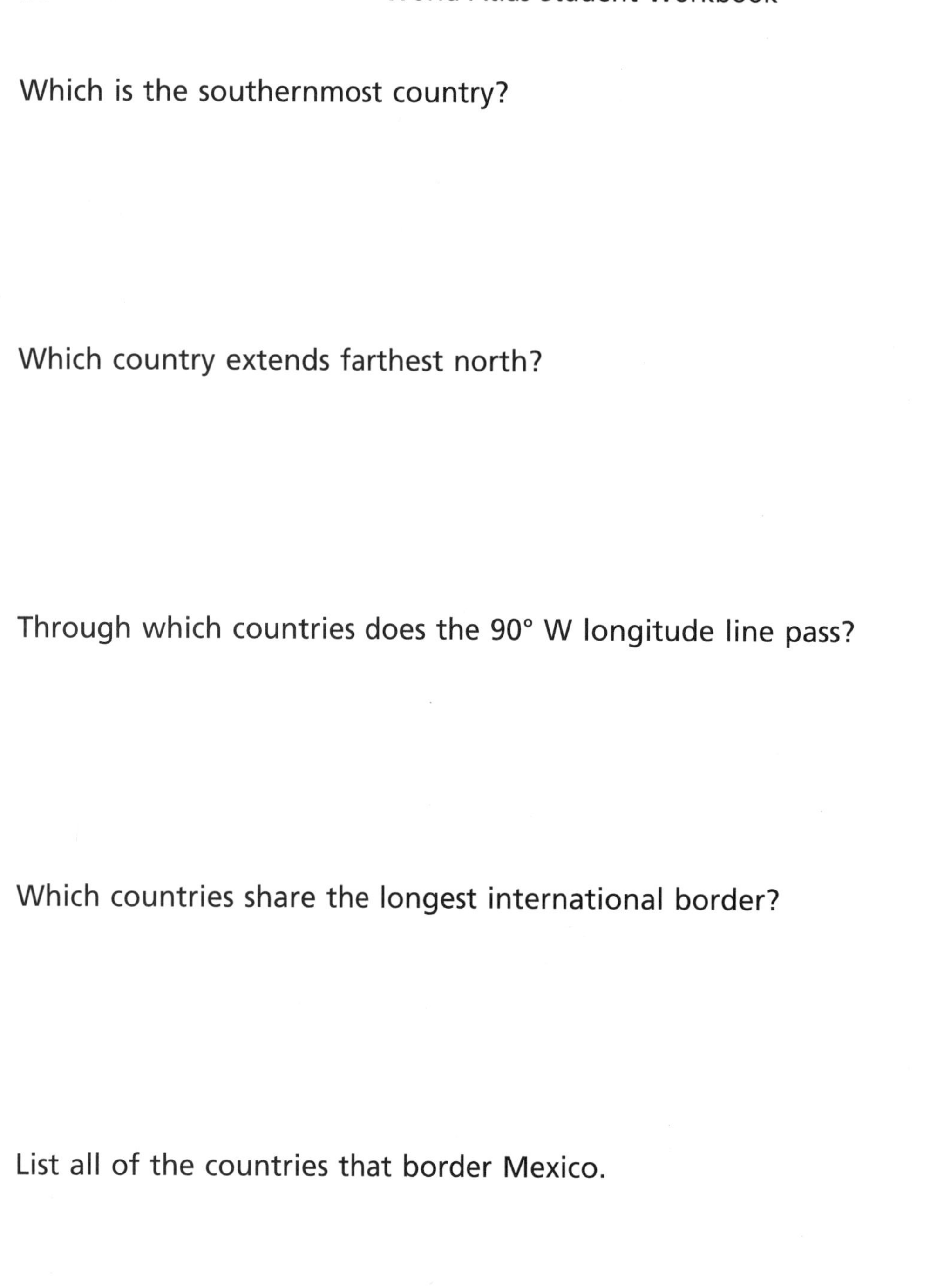

Which is the southernmost country?

Which country extends farthest north?

Through which countries does the 90° W longitude line pass?

Which countries share the longest international border?

List all of the countries that border Mexico.

Which countries have territory that extends above the Arctic Circle?

Chicago, Illinois, USA *Corbis Digital Stock*

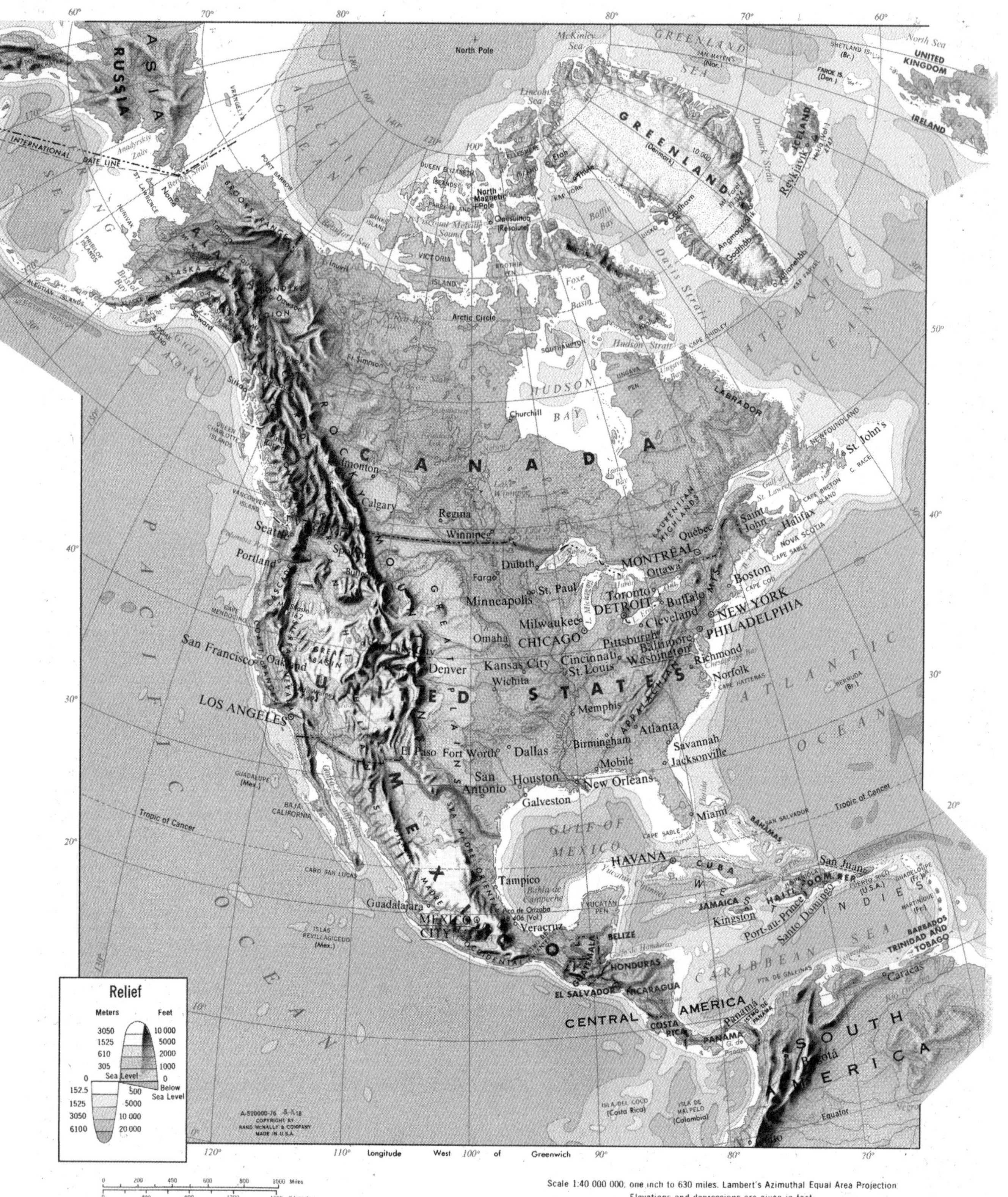
North Pole
ASIA
RUSSIA
INTERNATIONAL DATE LINE
GREENLAND
UNITED KINGDOM
IRELAND
ICELAND
Reykjavík
Davis Strait
Baffin Bay
Foxe Basin
Hudson Strait
HUDSON BAY
Churchill
LABRADOR
NEWFOUNDLAND
St. John's
Arctic Circle
ALASKA
Nome
Bristol Bay
ALEUTIAN ISLANDS
Seward
VICTORIA ISLAND
C A N A D A
Edmonton
Calgary
Regina
Winnipeg
VANCOUVER ISLAND
Seattle
Portland
Spokane
Duluth
Fargo
St. Paul
Minneapolis
MONTREAL
Ottawa
Québec
Saint John
Halifax
NOVA SCOTIA
Boston
NEW YORK
PHILADELPHIA
Toronto
DETROIT
Buffalo
Cleveland
Milwaukee
CHICAGO
Pittsburgh
Baltimore
Washington
Richmond
Norfolk
Omaha
Kansas City
Cincinnati
St. Louis
Denver
Wichita
San Francisco
Oakland
GREAT BASIN
LOS ANGELES
U N I T E D S T A T E S
Memphis
Atlanta
Birmingham
Savannah
Jacksonville
Mobile
New Orleans
El Paso
Fort Worth
Dallas
San Antonio
Houston
Galveston
Miami
GULF OF MEXICO
Tropic of Cancer
BAJA CALIFORNIA
HAVANA
CUBA
San Juan
JAMAICA
Kingston
Port-au-Prince
Santo Domingo
W E S T I N D I E S
BARBADOS
TRINIDAD AND TOBAGO
Caracas
Tampico
Guadalajara
MEXICO CITY
Veracruz
M E X I C O
BELIZE
HONDURAS
EL SALVADOR
NICARAGUA
CENTRAL AMERICA
COSTA RICA
PANAMA
SOUTH AMERICA
Bogotá
Equator
CARIBBEAN SEA
P A C I F I C O C E A N
A T L A N T I C O C E A N
Relief
Meters
Feet
3050
1525
610
305
0
152.5
1525
3050
6100
10 000
5000
2000
1000
Sea Level
0
500
Below Sea Level
5000
10 000
20 000
120°
110°
Longitude West 100° of Greenwich
90°
80°
70°
0 200 400 600 800 1000 Miles
0 400 800 1200 1600 Kilometers
Scale 1:40 000 000; one inch to 630 miles. Lambert's Azimuthal Equal Area Projection
Elevations and depressions are given in feet

## *North America Physical Map*

Name: ______________________ Date: ______________________

See page **89** in Goode's World Atlas (21e) for a full color map.

What are the major **mountain ranges** of the following countries?

Canada ______________________
United States ______________________
Mexico ______________________

What are the major **rivers** in each of these countries?

Canada ______________________
United States ______________________
Mexico ______________________

List the country(s) where you find each of the following **features**.

Laurentian Highlands ______________________
Sierra Madre Oriental ______________________
Cascade Range ______________________
Yucatan Peninsula ______________________
Great Basin ______________________
Great Bear Lake ______________________
Lake Superior ______________________
Sierra Madre Occidental ______________________
Great Salt Lake ______________________
Golfo de California ______________________
Mount McKinley ______________________
Rocky Mountains ______________________
Istmo dé Panama ______________________
Mount Whitney ______________________
Lake Ontario ______________________
Hudson Bay ______________________
Lago de Nicaragua ______________________
Mount Logan ______________________
Lake Huron ______________________
Ungava Peninsula ______________________
Volcan Pico de Orizaba ______________________

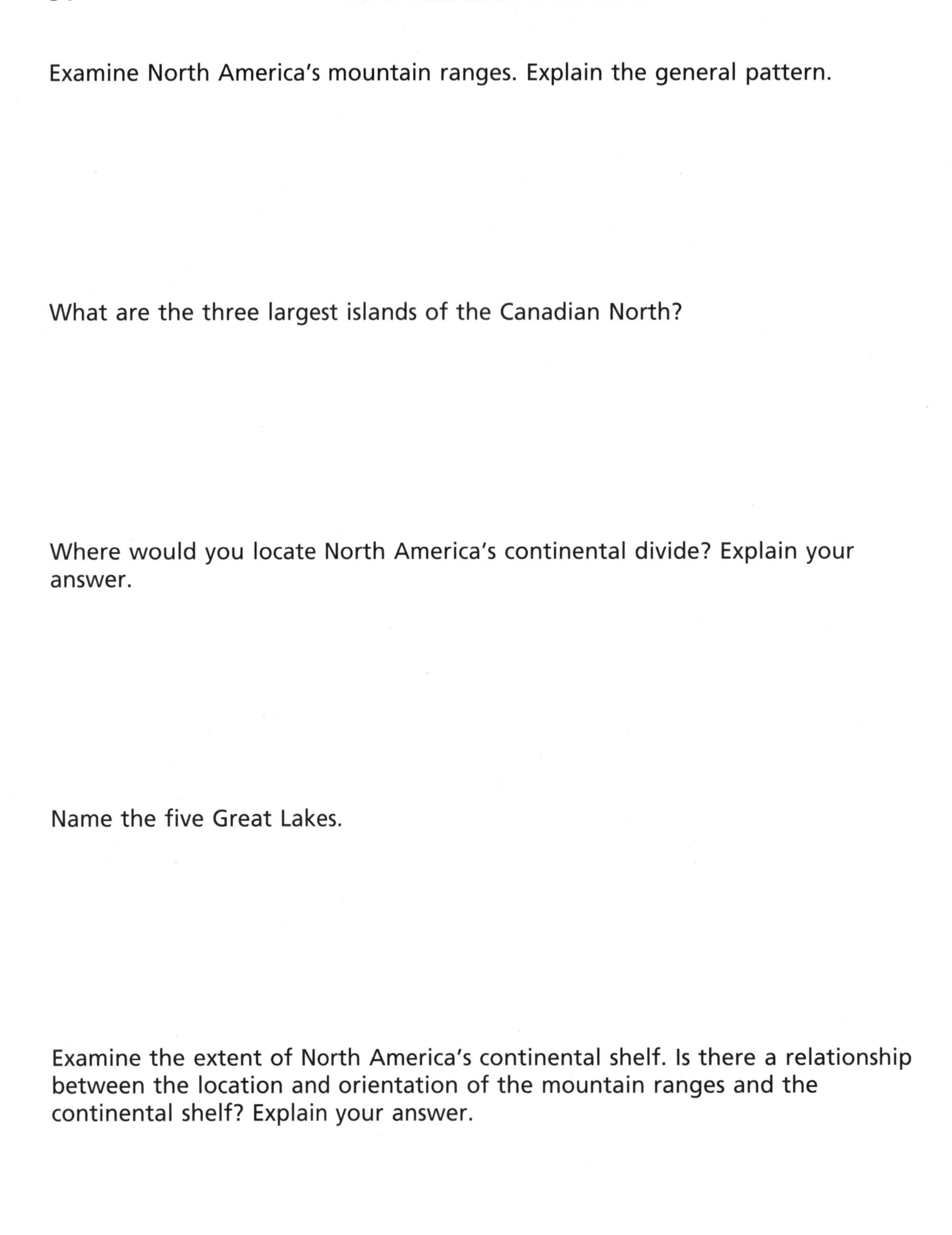

Examine North America's mountain ranges. Explain the general pattern.

What are the three largest islands of the Canadian North?

Where would you locate North America's continental divide? Explain your answer.

Name the five Great Lakes.

Examine the extent of North America's continental shelf. Is there a relationship between the location and orientation of the mountain ranges and the continental shelf? Explain your answer.

Chichen Itza, Mexico *Corbis Digital Stock*

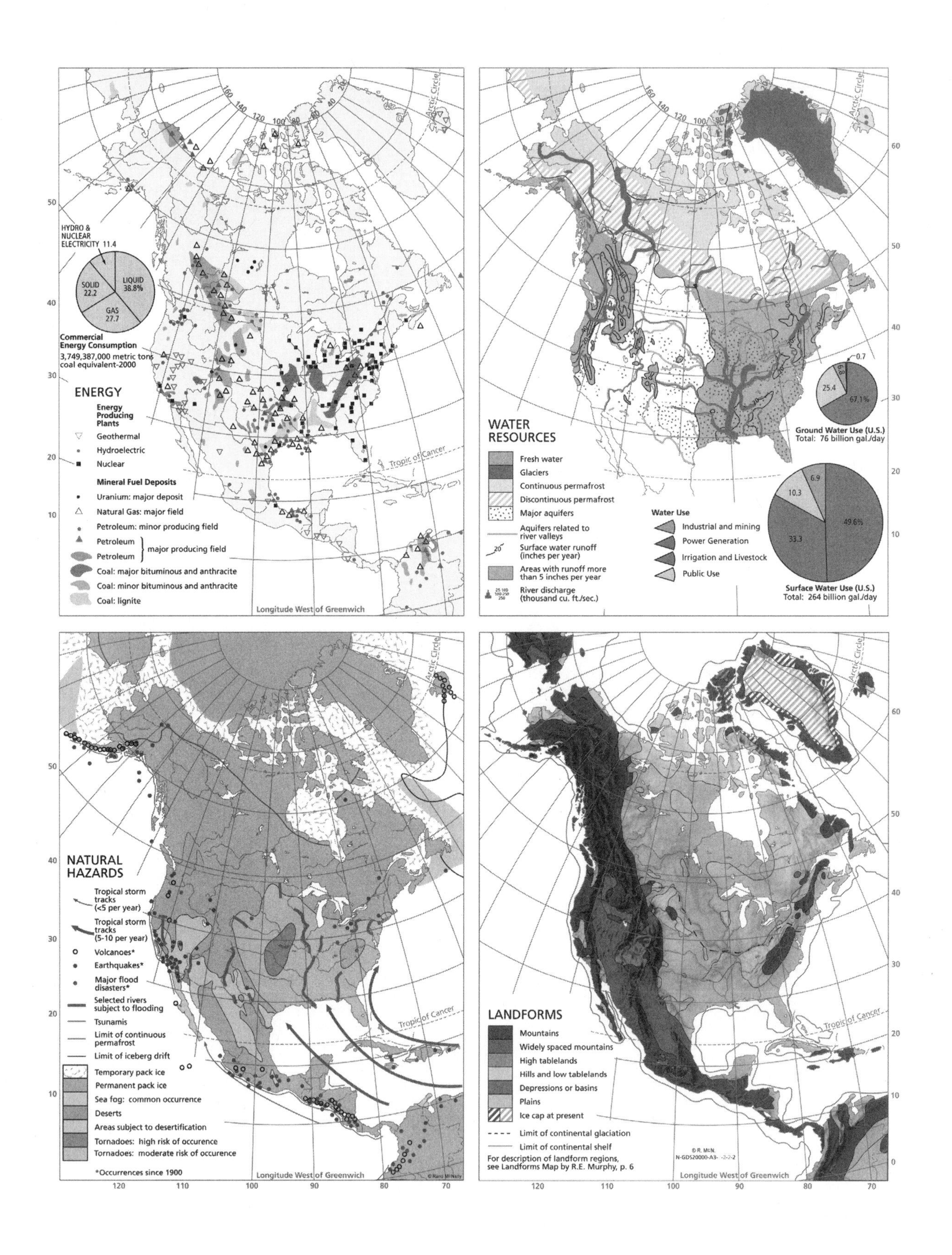
ENERGY
Energy Producing Plants
Geothermal
Hydroelectric
Nuclear
Mineral Fuel Deposits
Uranium: major deposit
Natural Gas: major field
Petroleum: minor producing field
Petroleum
Petroleum
major producing field
Coal: major bituminous and anthracite
Coal: minor bituminous and anthracite
Coal: lignite
HYDRO & NUCLEAR ELECTRICITY 11.4
SOLID 22.2
LIQUID 38.8%
GAS 27.7
Commercial Energy Consumption
3,749,387,000 metric tons coal equivalent-2000
Tropic of Cancer
Arctic Circle
Longitude West of Greenwich
WATER RESOURCES
Fresh water
Glaciers
Continuous permafrost
Discontinuous permafrost
Major aquifers
Aquifers related to river valleys
Surface water runoff (inches per year)
Areas with runoff more than 5 inches per year
River discharge (thousand cu. ft./sec.)
Water Use
Industrial and mining
Power Generation
Irrigation and Livestock
Public Use
Ground Water Use (U.S.)
Total: 76 billion gal./day
67.1%
25.4
0.7
Surface Water Use (U.S.)
Total: 264 billion gal./day
49.6%
33.3
10.3
6.9
NATURAL HAZARDS
Tropical storm tracks (<5 per year)
Tropical storm tracks (5-10 per year)
Volcanoes*
Earthquakes*
Major flood disasters*
Selected rivers subject to flooding
Tsunamis
Limit of continuous permafrost
Limit of iceberg drift
Temporary pack ice
Permanent pack ice
Sea fog: common occurrence
Deserts
Areas subject to desertification
Tornadoes: high risk of occurence
Tornadoes: moderate risk of occurence
*Occurrences since 1900
LANDFORMS
Mountains
Widely spaced mountains
High tablelands
Hills and low tablelands
Depressions or basins
Plains
Ice cap at present
Limit of continental glaciation
Limit of continental shelf
For description of landform regions, see Landforms Map by R.E. Murphy, p. 6
120
110
100
90
80
70

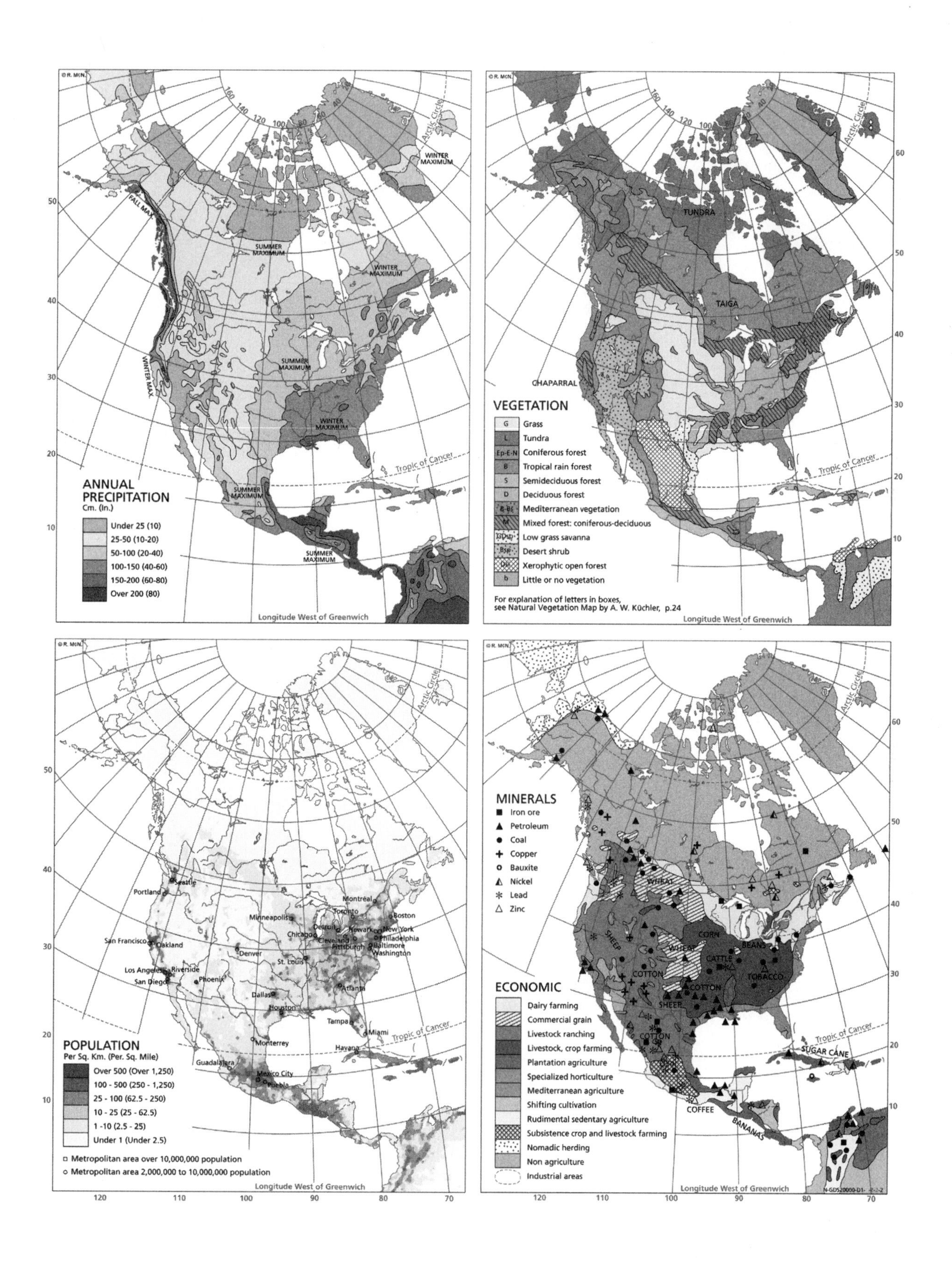

ANNUAL PRECIPITATION
Cm. (In.)
Under 25 (10)
25-50 (10-20)
50-100 (20-40)
100-150 (40-60)
150-200 (60-80)
Over 200 (80)
WINTER MAXIMUM
FALL MAX.
SUMMER MAXIMUM
WINTER MAX.
Arctic Circle
Tropic of Cancer
Longitude West of Greenwich
VEGETATION
Grass
Tundra
Coniferous forest
Tropical rain forest
Semideciduous forest
Deciduous forest
Mediterranean vegetation
Mixed forest: coniferous-deciduous
Low grass savanna
Desert shrub
Xerophytic open forest
Little or no vegetation
For explanation of letters in boxes, see Natural Vegetation Map by A. W. Küchler, p.24
TUNDRA
TAIGA
CHAPARRAL
POPULATION
Per Sq. Km. (Per. Sq. Mile)
Over 500 (Over 1,250)
100 - 500 (250 - 1,250)
25 - 100 (62.5 - 250)
10 - 25 (25 - 62.5)
1 -10 (2.5 - 25)
Under 1 (Under 2.5)
Metropolitan area over 10,000,000 population
Metropolitan area 2,000,000 to 10,000,000 population
Seattle
Portland
San Francisco
Oakland
Los Angeles
Riverside
San Diego
Phoenix
Denver
Minneapolis
Chicago
Detroit
Cleveland
Pittsburgh
St. Louis
Dallas
Houston
Atlanta
Tampa
Miami
Montréal
Toronto
Boston
New York
Newark
Philadelphia
Baltimore
Washington
Monterrey
Guadalajara
Mexico City
Puebla
Havana
MINERALS
Iron ore
Petroleum
Coal
Copper
Bauxite
Nickel
Lead
Zinc
ECONOMIC
Dairy farming
Commercial grain
Livestock ranching
Livestock, crop farming
Plantation agriculture
Specialized horticulture
Mediterranean agriculture
Shifting cultivation
Rudimental sedentary agriculture
Subsistence crop and livestock farming
Nomadic herding
Non agriculture
Industrial areas
WHEAT
SHEEP
CORN
BEANS
CATTLE
COTTON
TOBACCO
SUGAR CANE
COFFEE
BANANAS
120 110 100 90 80 70

Banff National Park, Alberta, Canada *Corbis Digital Stock*

## *North America Thematic Maps*

Name: ________________________________ Date: ______________________________

See pages **67-68** in Goode's World Atlas (21e) for full color maps.

---

What are the major natural hazards in North America? Do they exhibit distinct geographic patterns? Explain your answer.

What are Mexico's major economic activities? Where in the country do you find specialized horticulture?

Where do the greatest amounts of annual precipitation occur on the continent? To what can you attribute this pattern?

Which North American country is most sparsely populated? To what can you attribute this status?

Which vegetative regime covers the largest areal extent in North America? In which country(s) is it located?

Where are the most arid regions of North America? How can you explain this pattern?

Examine North America's population distribution. Where are the most densely populated regions? How can you explain these concentrations? What factors can help to explain the vast regions of sparse population?

Where are the major petroleum fields of North America? Are these associated with any particular landform regions?

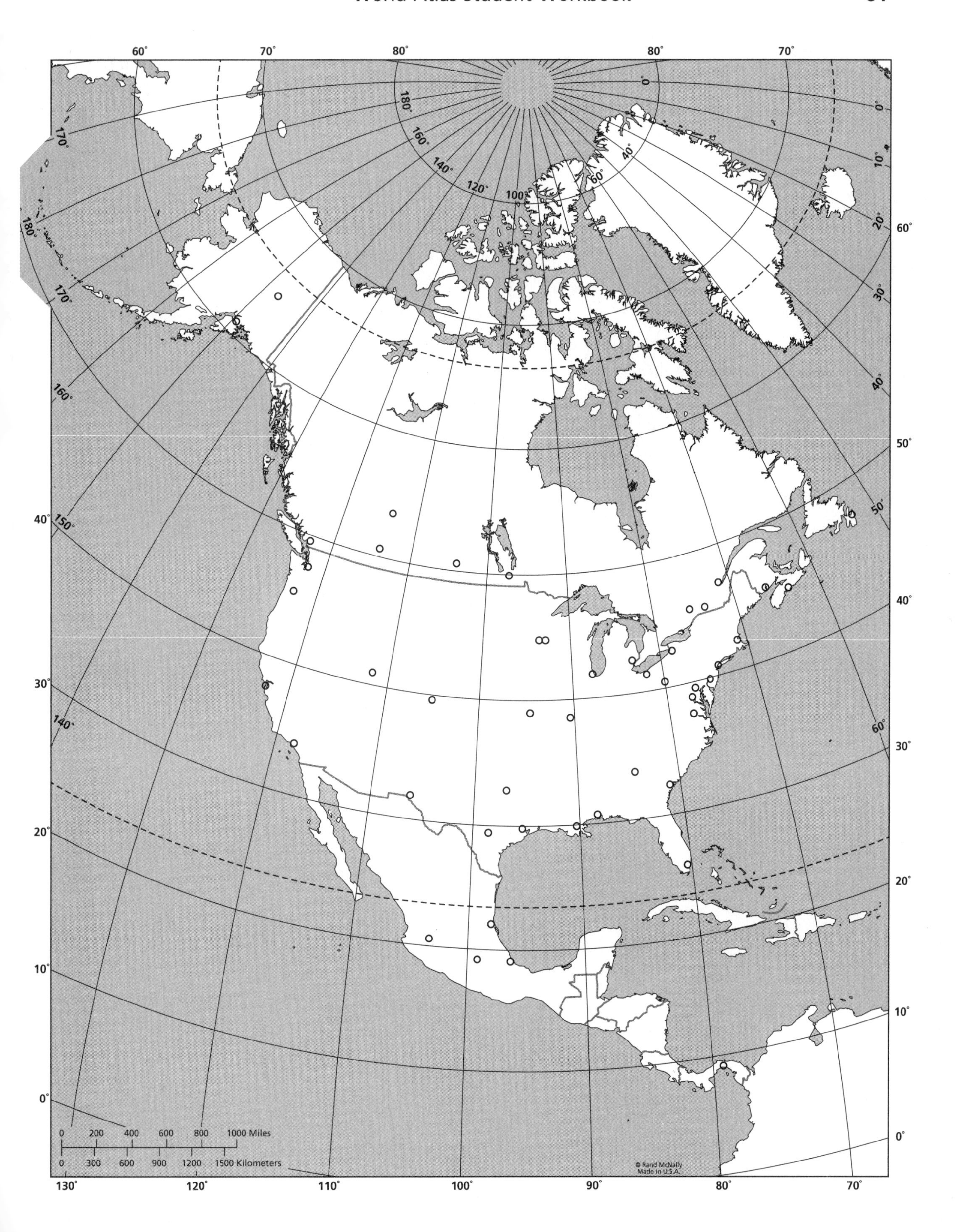
60°
70°
80°
80°
70°
0°
180°
170°
160°
140°
120°
100°
80°
60°
40°
10°
20°
30°
40°
50°
60°
180°
170°
160°
150°
140°
60°
50°
40°
30°
20°
10°
0°
40°
30°
20°
10°
0°
0 200 400 600 800 1000 Miles
0 300 600 900 1200 1500 Kilometers
© Rand McNally
Made in U.S.A.
130°
120°
110°
100°
90°
80°
70°

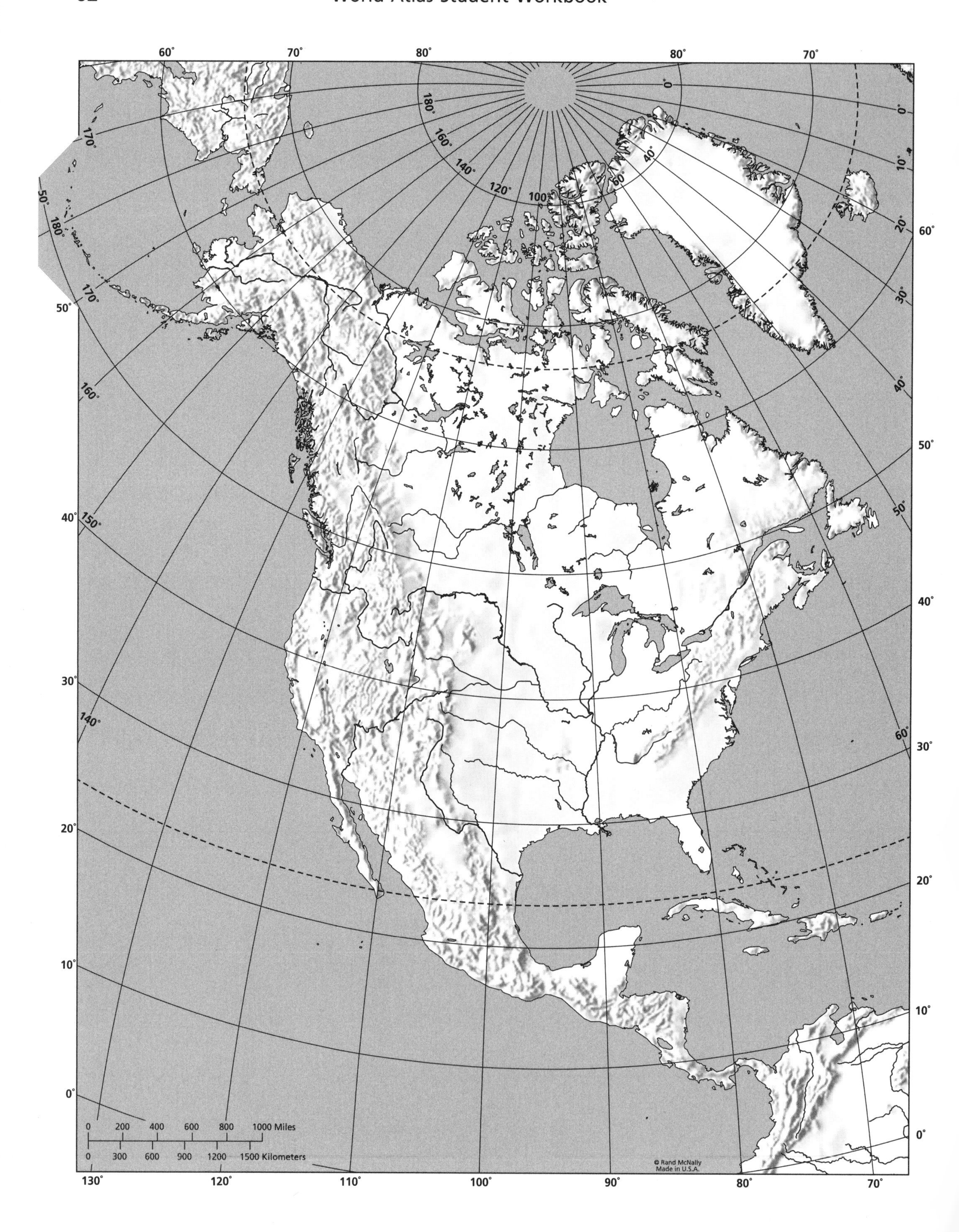
60°
70°
80°
80°
70°
0°
180°
160°
140°
120°
100°
60°
40°
0°
10°
20°
30°
40°
50°
60°
170°
50°
180°
170°
160°
150°
140°
60°
50°
40°
30°
20°
10°
0°
50°
40°
30°
20°
10°
0°
0 200 400 600 800 1000 Miles
0 300 600 900 1200 1500 Kilometers
© Rand McNally
Made in U.S.A.
130°
120°
110°
100°
90°
80°
70°

## ***North America Blank Maps*** *(Political / Physical)*

Name: ______________________ Date: ______________________

See page(s) **88-89** in Goode's World Atlas (21e) for full color maps.

### Political Map

Label the following **countries** on the blank political map of North America.

| | |
|---|---|
| Belize<br>Canada<br>Costa Rica<br>El Salvador<br>Guatemala | Honduras<br>Mexico<br>Nicaragua<br>Panama<br>United States |

Label the following **cities** on the blank political map of North America.

| | |
|---|---|
| Anchorage<br>Atlanta<br>Baltimore<br>Boston<br>Buffalo<br>Calgary<br>Chicago<br>Cleveland<br>Dallas<br>Denver<br>Detroit<br>Edmonton<br>El Paso<br>Fairbanks<br>Guadalajara<br>Halifax<br>Houston<br>Kansas city<br>Los Angeles<br>Mexico City<br>Miami<br>Minneapolis<br>Mobile<br>Montreal | New Orleans<br>New York<br>Ottawa<br>Panama<br>Philadelphia<br>Pittsburgh<br>Portland<br>Quebec<br>Regina<br>Richmond<br>Saint John<br>Salt Lake City<br>San Antonio<br>San Francisco<br>Savannah<br>Seattle<br>ST. John's<br>ST. Louis<br>ST. Paul<br>Tampico<br>Vancouver<br>Veracuz<br>Washington, D.C.<br>Winnipeg |

## Physical Map

Label the following **features** on the blank physical map of North America.

### *Mountain Ranges*

| | |
|---|---|
| Appalachian<br>Alaska<br>Cascade<br>Coast | Rocky<br>Sierra Madre Occidental<br>Sierra Madre Oriental |

### *Rivers*

| | |
|---|---|
| Arkansas<br>Colorado<br>Columbia<br>Mackenzie<br>Mississippi<br>Missouri<br>Ohio | Peace<br>Platte<br>Red<br>Rio Grande<br>Saint Lawrence<br>Snake<br>Yukon |

### *Bodies of Water*

| | |
|---|---|
| Arctic Ocean<br>Atlantic Ocean<br>Bay of Fundy<br>Beaufort Sea<br>Bering Sea<br>Davis Strait<br>Golfo de California<br>Great Bear Lake<br>Gulf of Alaska<br>Gulf of Mexico | Gulf of ST. Lawrence<br>Hudson Bay<br>Hudson Strait<br>Lake Erie<br>Lake Huron<br>Lake Michigan<br>Lake Ontario<br>Lake Superior<br>Lake Winnipeg<br>Pacific Ocean |

### *Islands*

| | |
|---|---|
| Aleutian<br>Banks<br>Cape Breton<br>Pribilof | Queen Charlotte<br>Queen Elizabeth<br>Vancouver<br>Victoria |

Rio de Janeiro, Brazil *Corbis Digital Stock*

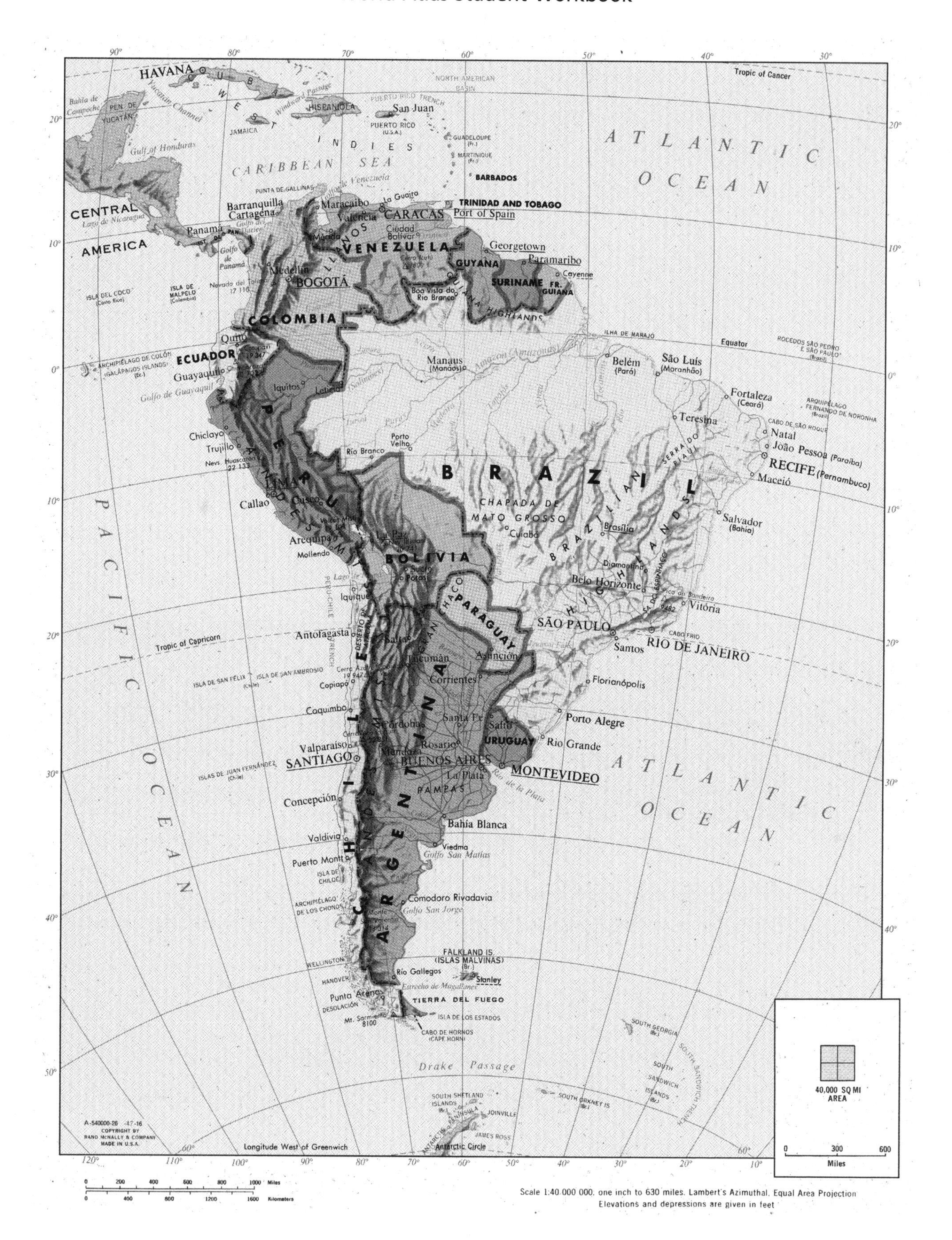


Scale 1:40,000,000; one inch to 630 miles. Lambert's Azimuthal, Equal Area Projection
Elevations and depressions are given in feet

## *South America Political Map*

Name: ______________________ Date: ______________________

See page **138** in Goode's World Atlas (21e) for a full color map.

List a **city** for each of the following letters.

| | |
|---|---|
| A ______________ | M ______________ |
| B ______________ | N ______________ |
| C ______________ | P ______________ |
| D ______________ | Q ______________ |
| F ______________ | R ______________ |
| G ______________ | S ______________ |
| I ______________ | T ______________ |
| J ______________ | V ______________ |
| L ______________ | |

List all of the countries through which the Tropic of Capricorn passes.

List all of the countries that are situated north of the Equator.

Which countries are located between the Equator and the Tropic of Capricorn?

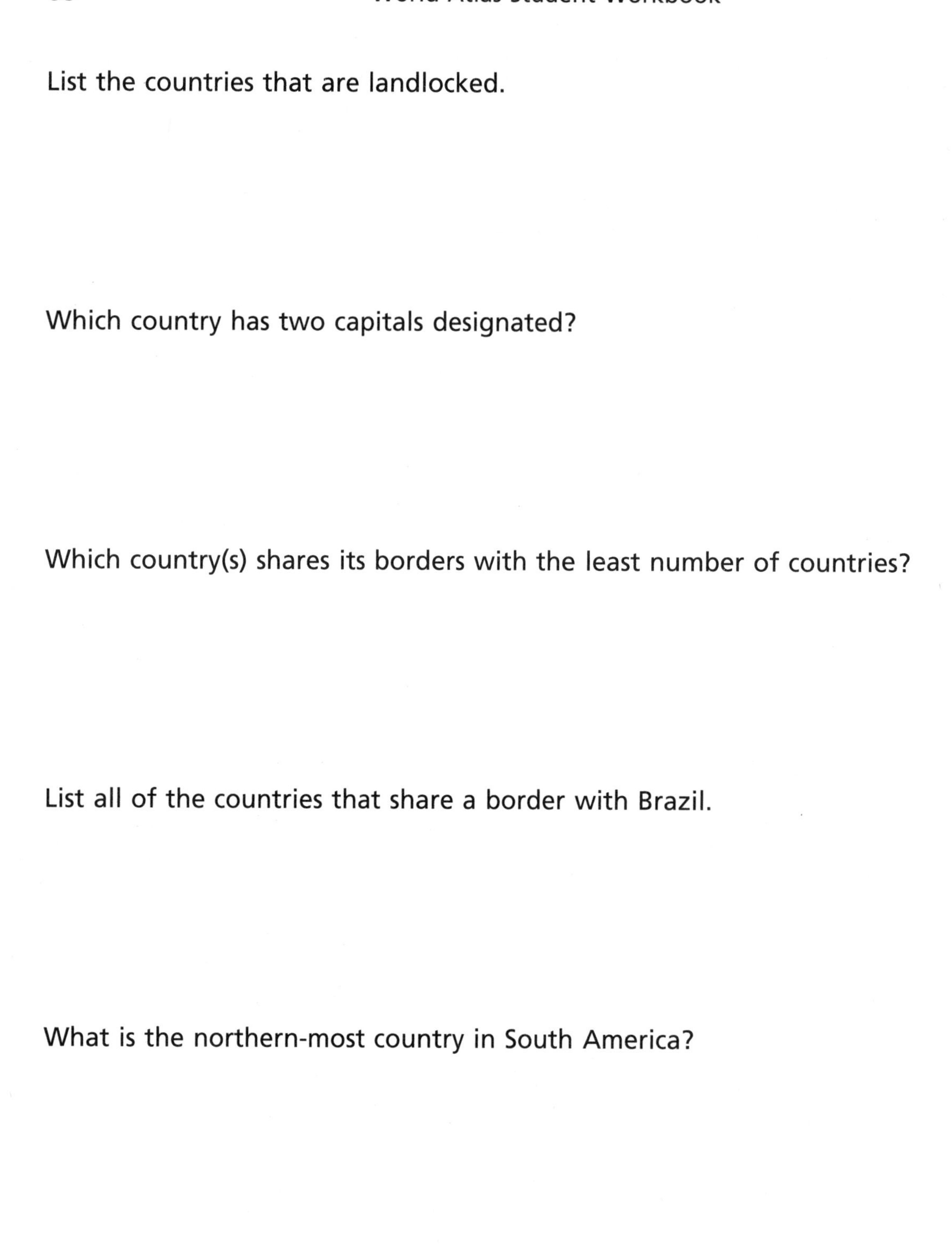

List the countries that are landlocked.

Which country has two capitals designated?

Which country(s) shares its borders with the least number of countries?

List all of the countries that share a border with Brazil.

What is the northern-most country in South America?

Which countries border the Caribbean Sea?

Guyana *PhotoDisc, Inc./Getty Images*

Scale 1:40 000 000; one inch to 630 miles. Lambert's Azimuthal, Equal Area Projection
Elevations and depressions are given in feet

## *South America Physical Map*

Name: ______________________ Date: ______________________

See page **139** in Goode's World Atlas (21e) for a full color map.

What are the major **mountain ranges** or **highlands** on the continent?

______________________

______________________

______________________

What are the major **rivers** on the continent?

______________________

______________________

______________________

List the country(s) where you find each of the following **features**.

Gran Chaco ______________________
Pampas ______________________
Archipiélago de los Chonos ______________________
Orinoco River ______________________
Lago Titicaca ______________________
Llanos ______________________
Paraná River ______________________
Desierto de Atacama ______________________
Cerro Aconcagua ______________________
Guianna Highlands ______________________
Amazon River ______________________
Andes Mountains ______________________
Volcán Misti ______________________
Golfo San Matias ______________________
Lago Maracaibo ______________________
Chapada de Mato Grosso ______________________
Magdalena River ______________________
Golfo de Guayaquil ______________________
Rio de la Plata ______________________
Nev Illimani ______________________
Ilha de Marajó ______________________
Iguassú Falls ______________________

What factors help to explain the location of the Atacama Desert?

Describe the continent's patterns of relief. How might you explain the location and orientation of the Andes Mountains?

What are the major barriers to overland travel across the continent? Which regions of the continent are more inaccessible than others? Explain your answer.

Where do you find vast areas of tropical rainforest? Why?

What types of hazards would ships encounter passing around Cabo de Hornos (Cape Horn)? Explain your answer.

Machu Picchu, Peru *Corbis Digital Stock*

HYDRO & NUCLEAR ELECTRICITY 16.2
SOLID 7.0
LIQUID 48.2%
GAS 28.6
Commercial Energy Consumption
434,205,000 metric tons coal equivalent-2000
ENERGY
Energy Producing Plants
Hydroelectric
Nuclear
Mineral Fuel Deposits
Uranium: major deposit
Natural Gas: major field
Petroleum: minor producing field
Petroleum
Petroleum
major producing field
Coal: minor bituminous
Coal: lignite
Longitude West of Greenwich
Equator
Tropic of Capricorn
SPANISH
CARIBAN
CHIBCHAN
ARAWAKAN
ARAWAKAN
TUPIAN
TUPIAN
QUECHUAN
AYMARAN
PORTUGUESE
GUARANI
GERMAN
SPANISH
GERMAN
PEOPLES
Predominant Racial Groups
European
Andean Indian
Other Indian
Mixed European and Indian
Mixed with large African proportion
Names on map represent significant language/culture groups
Map after Preston E. James
NATURAL HAZARDS
Volcanoes*
Earthquakes*
Major flood disasters*
Tsunami
Limit of iceberg drifts
Deserts
Areas subject to desertification
*Occurrences since 1900
LANDFORMS
Mountains
Widely spaced mountains
High tablelands
Hills and low tablelands
Depressions or basins
Plains
Limit of continental shelf
For description of landform regions, see Landforms Map by R.E. Murphy, p.6
© Rand McNally

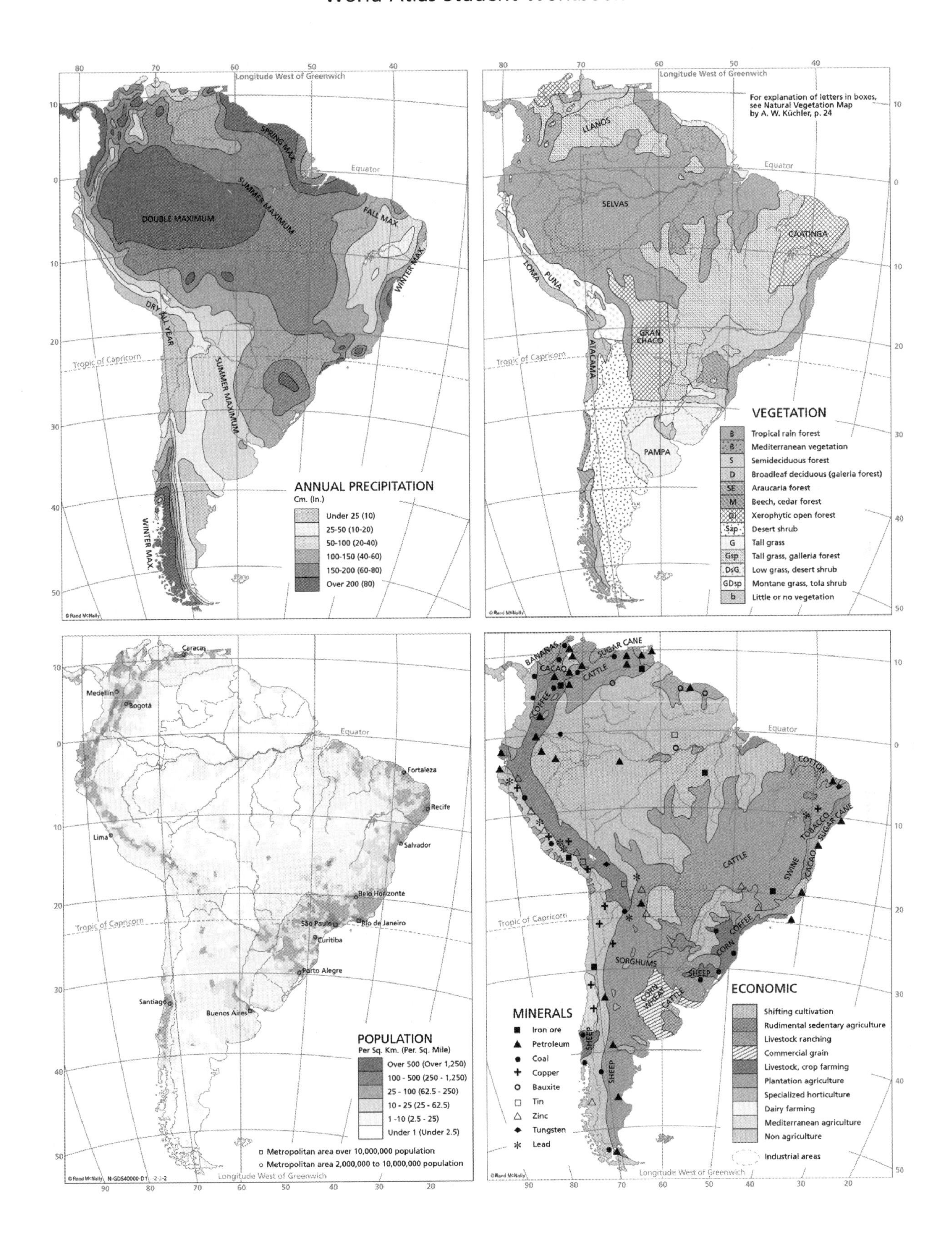
ANNUAL PRECIPITATION
Cm. (In.)
Under 25 (10)
25-50 (10-20)
50-100 (20-40)
100-150 (40-60)
150-200 (60-80)
Over 200 (80)
DOUBLE MAXIMUM
SPRING MAX.
SUMMER MAXIMUM
FALL MAX.
WINTER MAX.
DRY ALL YEAR
Longitude West of Greenwich
Equator
Tropic of Capricorn
VEGETATION
B Tropical rain forest
B Mediterranean vegetation
S Semideciduous forest
D Broadleaf deciduous (galeria forest)
SE Araucaria forest
M Beech, cedar forest
Di Xerophytic open forest
Sap Desert shrub
G Tall grass
Gsp Tall grass, galleria forest
DsG Low grass, desert shrub
GDsp Montane grass, tola shrub
b Little or no vegetation
For explanation of letters in boxes, see Natural Vegetation Map by A. W. Küchler, p. 24
LLANOS
SELVAS
CAATINGA
LOMA
PUNA
GRAN CHACO
ATACAMA
PAMPA
POPULATION
Per Sq. Km. (Per. Sq. Mile)
Over 500 (Over 1,250)
100 - 500 (250 - 1,250)
25 - 100 (62.5 - 250)
10 - 25 (25 - 62.5)
1 -10 (2.5 - 25)
Under 1 (Under 2.5)
Metropolitan area over 10,000,000 population
Metropolitan area 2,000,000 to 10,000,000 population
Caracas
Medellín
Bogotá
Fortaleza
Recife
Lima
Salvador
Belo Horizonte
São Paulo
Rio de Janeiro
Curitiba
Porto Alegre
Santiago
Buenos Aires
ECONOMIC
Shifting cultivation
Rudimental sedentary agriculture
Livestock ranching
Commercial grain
Livestock, crop farming
Plantation agriculture
Specialized horticulture
Dairy farming
Mediterranean agriculture
Non agriculture
Industrial areas
MINERALS
Iron ore
Petroleum
Coal
Copper
Bauxite
Tin
Zinc
Tungsten
Lead
BANANAS
CACAO
SUGAR CANE
CATTLE
COFFEE
COTTON
TOBACCO
SWINE
CORN
SORGHUMS
SHEEP
WHEAT
© Rand McNally N-GDS40000-D1

Panama Canal *PhotoDisc, Inc.*

## *South America Thematic Maps*

Name: ______________________________ Date: ____________________________

See pages **136-137** in Goode's World Atlas (21e) for full color maps.

---

Examine the map of South America's natural hazards. Describe the distribution of volcanoes and earthquakes on the continent. To what can you attribute this pattern?

Examine South America's population distribution. What pattern emerges? What factors may have contributed to this spatial pattern?

Where does shifting cultivation occur on the continent? Why is it prevalent in this particular region?

What regions experience the highest amounts of annual precipitation on the continent? Where do the least amounts occur? What factors contribute to this pattern?

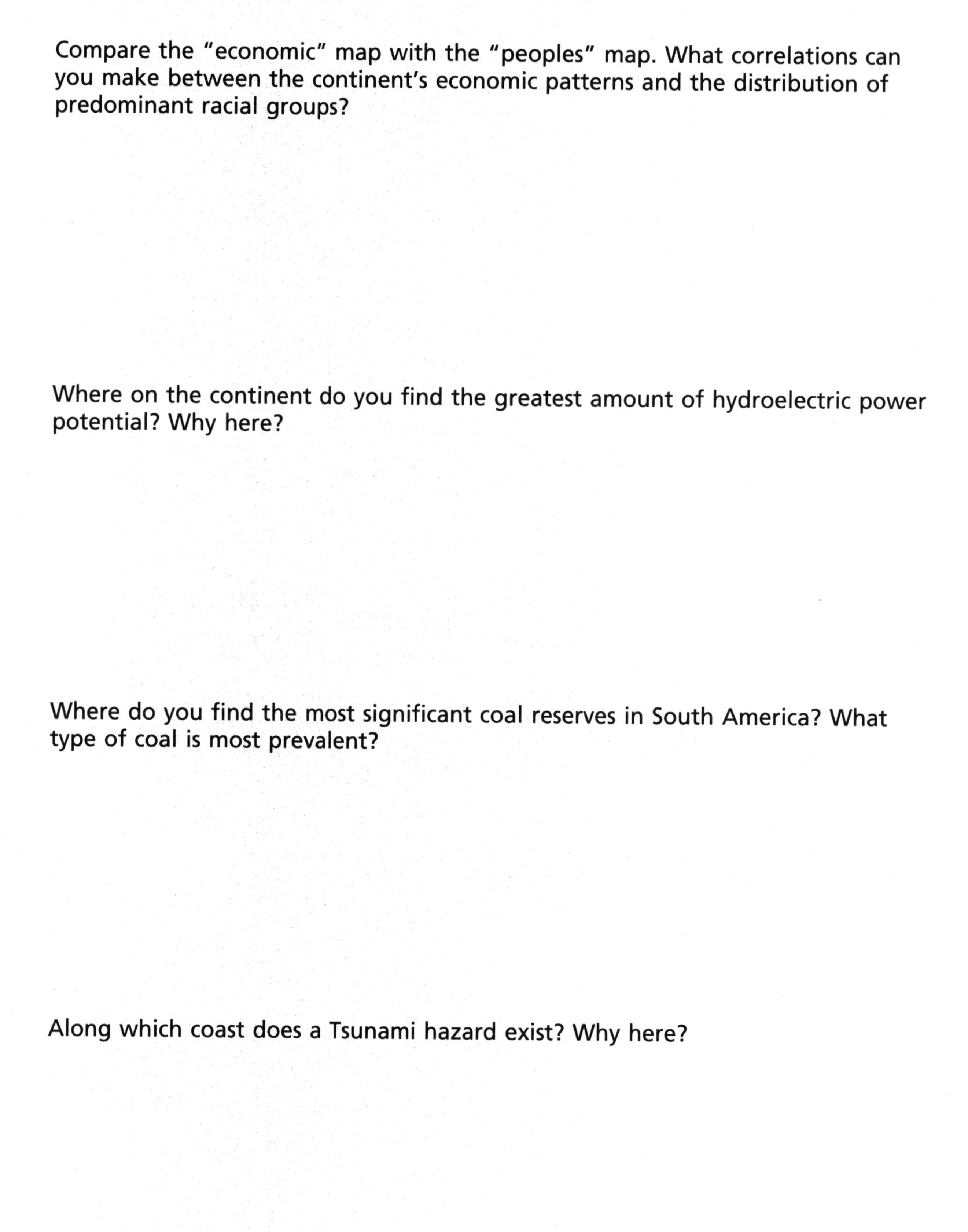

Compare the "economic" map with the "peoples" map. What correlations can you make between the continent's economic patterns and the distribution of predominant racial groups?

Where on the continent do you find the greatest amount of hydroelectric power potential? Why here?

Where do you find the most significant coal reserves in South America? What type of coal is most prevalent?

Along which coast does a Tsunami hazard exist? Why here?

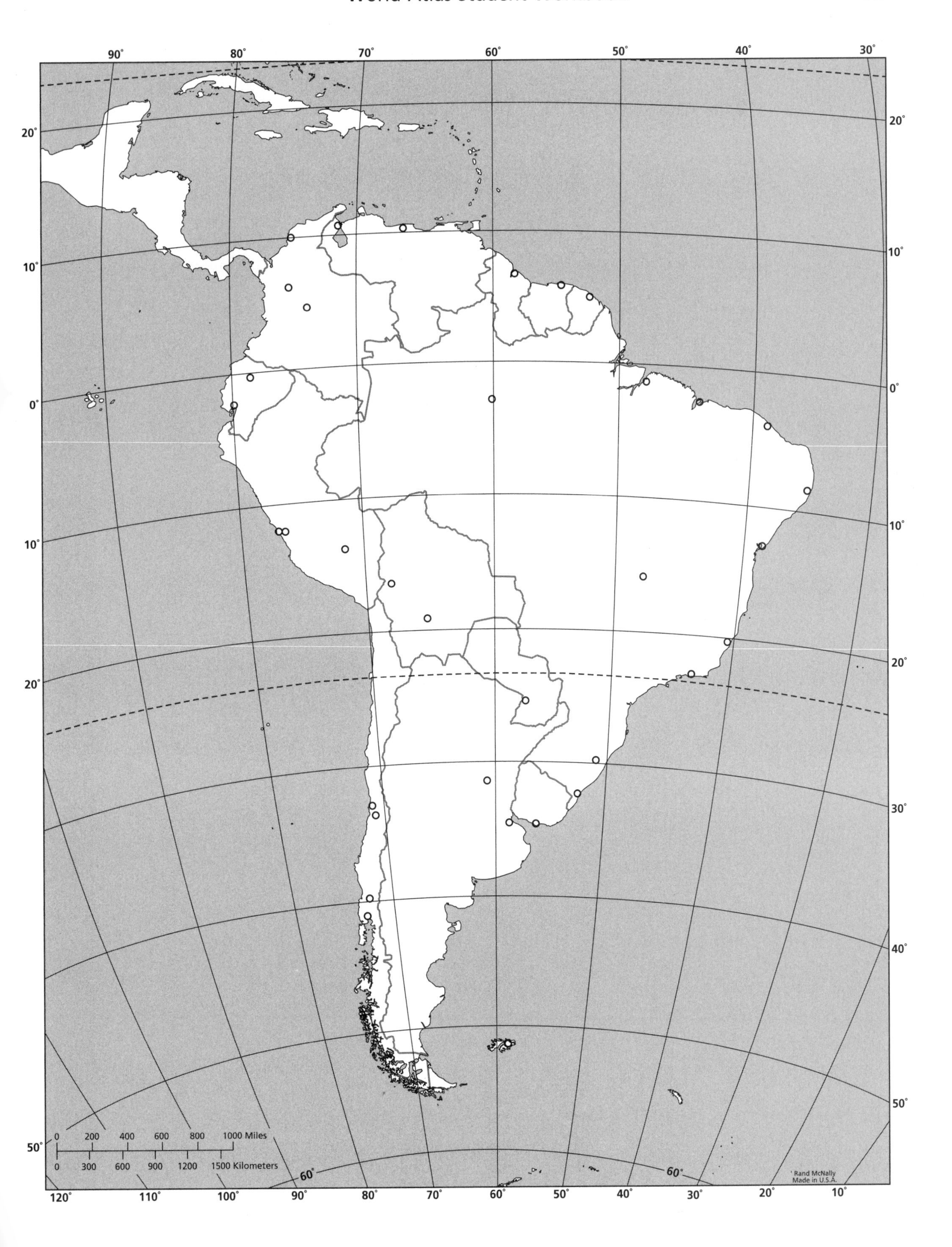
90°
80°
70°
60°
50°
40°
30°
20°
10°
0°
10°
20°
20°
10°
0°
10°
20°
30°
40°
50°
50°
0 200 400 600 800 1000 Miles
0 300 600 900 1200 1500 Kilometers
60°
60°
120°
110°
100°
90°
80°
70°
60°
50°
40°
30°
20°
10°
Rand McNally
Made in U.S.A.

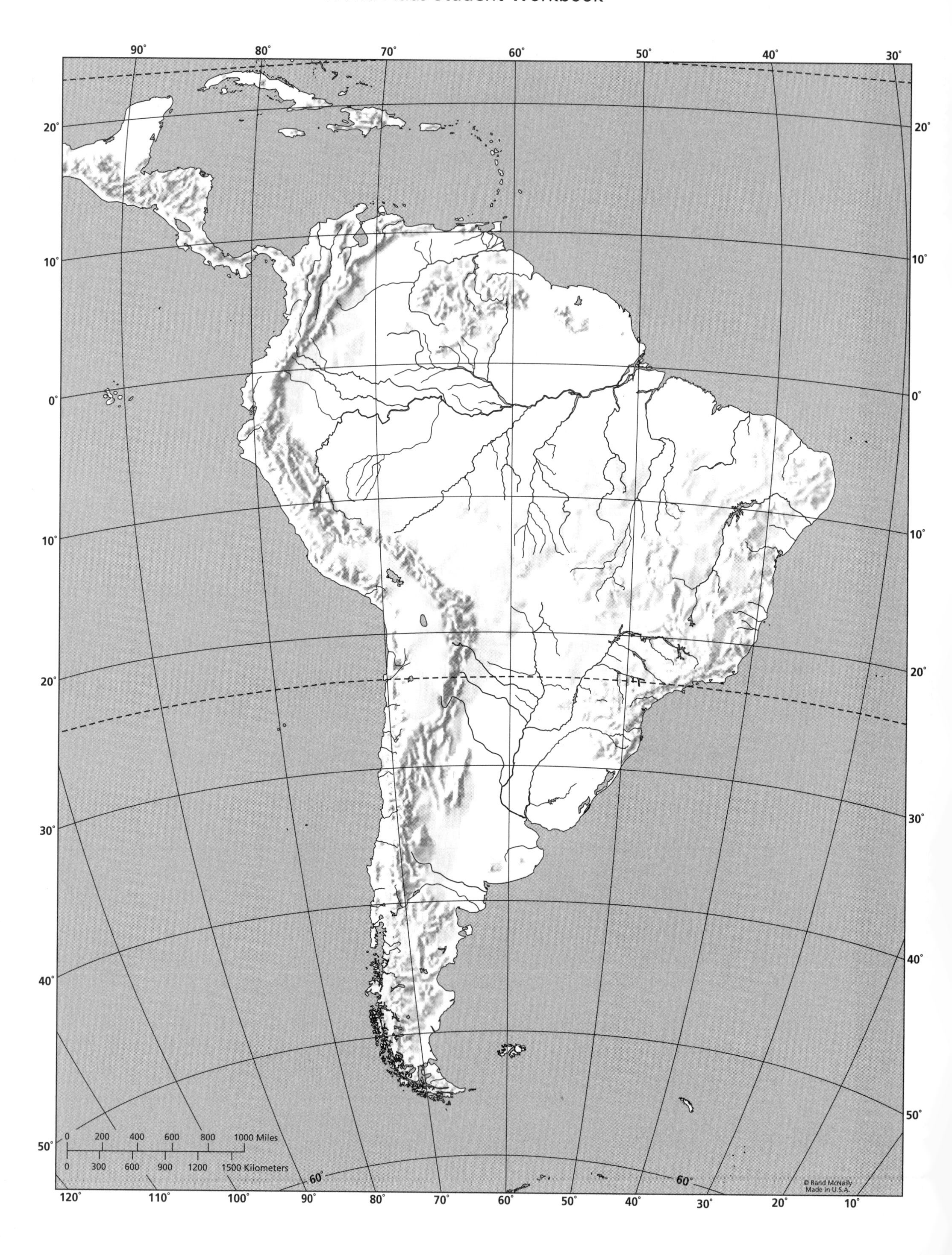
90°
80°
70°
60°
50°
40°
30°
20°
10°
0°
10°
20°
30°
40°
50°
0 200 400 600 800 1000 Miles
0 300 600 900 1200 1500 Kilometers
60°
60°
120°
110°
100°
90°
80°
70°
60°
50°
40°
30°
20°
10°
© Rand McNally
Made in U.S.A.

## *South America Blank Maps (Political / Physical)*

Name: ______________________ Date: ______________________

See pages **138-139** in Goode's World Atlas (21e) for full color maps.

### Political Map

Label the following **countries** on the blank political map of South America.

| | |
|---|---|
| Argentina | Guyana |
| Bolivia | Paraguay |
| Brazil | Peru |
| Chile | Suriname |
| Colombia | Uruguay |
| Ecuador | Venezuela |
| French Guiana | |

Label the following **cities** on the blank political map of South America.

| | |
|---|---|
| Asunción | Montevideo |
| Belem | Paramaribo |
| Bogotá | Porto Alegre |
| Brasilia | Puerto Montt |
| Buenos Aires | Quito |
| Callao | Recife |
| Caracas | Rio de Janeiro |
| Cartagena | Rio Grande |
| Cayenne | Salvador |
| Cusco | Sante Fe |
| Fortaleza | Santiago |
| Georgetown | Saõ Luis |
| Guayaquil | Stanley |
| La Paz | Sucre |
| Lima | Valdivia |
| Manaus | Valparaiso |
| Maricaibo | Vitoria |
| Medellin | |

## Physical Map

Label the following **features** on the blank physical map of South America.

### *Mountain Ranges and Highlands*

| | |
|---|---|
| Andes<br>Brazilian Highlands | Chapada de Mato Grosso<br>Guiana Highlands |

### *Rivers*

| | |
|---|---|
| Amazon<br>Bermejo<br>Branco<br>Cauca<br>Colorado<br>Juruá<br>Madeira<br>Magdalena<br>Negro | Orinoco<br>Paraguay<br>Paraná<br>Salado<br>Saõ Francisco<br>Tapajos<br>Tocantins<br>Uruguay<br>Xingú |

### *Bodies of Water*

| | |
|---|---|
| Atlantic Ocean<br>Caribbean Sea<br>Drake Passage<br>Golfo de Guayaquil<br>Golfo de Panama<br>Golfo de Venezuela | Golfo San Jorge<br>Golfo San Matias<br>Lago de Poopó<br>Lago Titicaca<br>Pacific Ocean |

### *Islands*

| | |
|---|---|
| Arquipélago Fernando de Noronha<br>Falkland (Malvinas)<br>Galápagos | Islas de Juan Fernández<br>Isla de Malpelo<br>Isla de San Félix |

Leaning Tower of Pisa, Italy *Corbis Digital Stock*

Scale 1: 16 000 000; one inch to 250 miles. Conic Projection
Elevations and depressions are given in feet

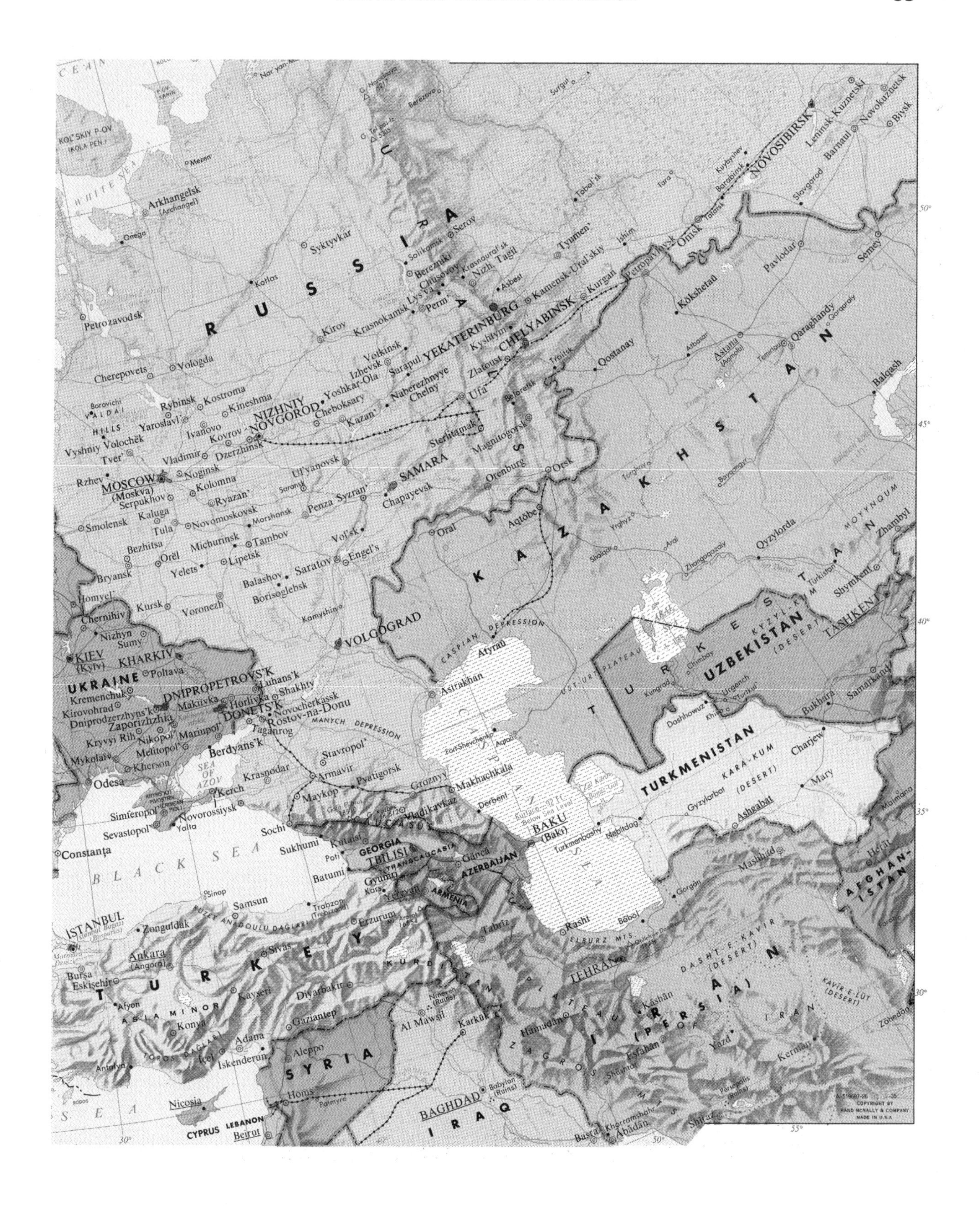
RUSSIA
URALS
KAZAKHSTAN
UZBEKISTAN
TURKMENISTAN
TURKESTAN
UKRAINE
GEORGIA
ARMENIA
AZERBAIJAN
TURKEY
SYRIA
IRAQ
IRAN (PERSIA)
AFGHANISTAN
CYPRUS
LEBANON
WHITE SEA
BLACK SEA
SEA OF AZOV
CASPIAN SEA
CASPIAN DEPRESSION
MANYCH DEPRESSION
KYZYL-KUM (DESERT)
KARA-KUM (DESERT)
DASHT-E-KAVIR (DESERT)
KAVIR-E-LUT (DESERT)
ASIA MINOR
ELBURZ MTS.
MOSCOW (Moskva)
KIEV (Kyiv)
KHARKIV
DNIPROPETROVS'K
DONETS'K
NIZHNIY NOVGOROD
YEKATERINBURG
CHELYABINSK
NOVOSIBIRSK
SAMARA
VOLGOGRAD
TASHKENT
TBILISI
BAKU (Baki)
TEHRAN
BAGHDAD
ISTANBUL
Ankara
Ashgabat
Astana (Aqmola)
Arkhangelsk (Archangel)
Petrozavodsk
Vologda
Yaroslavl'
Kazan'
Perm'
Ufa
Omsk
Orenburg
Saratov
Astrakhan
Atyrau
Rostov-na-Donu
Odesa
Sevastopol'
Simferopol'
Novorossiysk
Krasnodar
Stavropol'
Groznyy
Makhachkala
Derbent
Batumi
Sukhumi
Samsun
Trabzon
Erzurum
Sivas
Kayseri
Konya
Adana
Iskenderun
Aleppo
Homs
Nicosia
Beirut
Al Mawsil
Kirkuk
Tabriz
Rasht
Hamadan
Esfahan
Kerman
Yazd
Mashhad
Herat
Bukhara
Samarkand
Shymkent
Qyzylorda
Aral
Aqtobe
Oral
Qostanay
Kokshetau
Pavlodar
Semey
Barnaul
Balqash
Qaraghandy
Turkmenbashy
Chardzhew
Mary
Basra
Abadan
Shiraz

Parthenon, Athens, Greece *Corbis Digital Stock*

## *Europe Political Map*

Name: ______________________ Date: ______________________

See pages **154-155** in Goode's World Atlas (21e) for full color maps.

List a **country** or major **city** for each of the following letters.

| | |
|---|---|
| A ______________________ | N ______________________ |
| B ______________________ | O ______________________ |
| C ______________________ | P ______________________ |
| D ______________________ | Q ______________________ |
| E ______________________ | R ______________________ |
| F ______________________ | S ______________________ |
| G ______________________ | T ______________________ |
| H ______________________ | U ______________________ |
| I ______________________ | V ______________________ |
| J ______________________ | W ______________________ |
| K ______________________ | Y ______________________ |
| L ______________________ | Z ______________________ |
| M ______________________ | |

List all the European countries that border the Mediterranean Sea.

List all the countries that border the Baltic Sea.

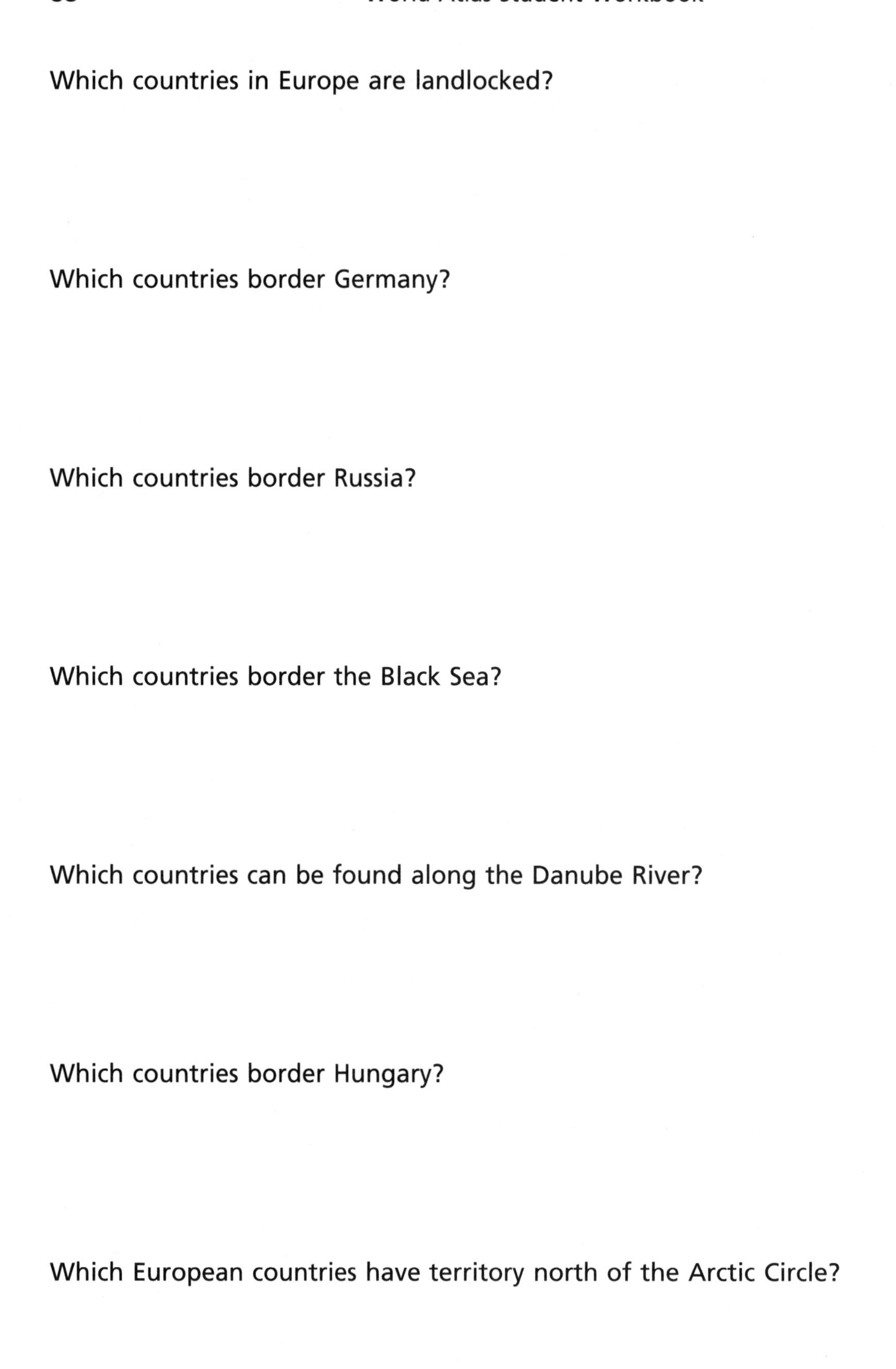

Which countries in Europe are landlocked?

Which countries border Germany?

Which countries border Russia?

Which countries border the Black Sea?

Which countries can be found along the Danube River?

Which countries border Hungary?

Which European countries have territory north of the Arctic Circle?

Matterhorn, Switzerland *Corbis Digital Stock*

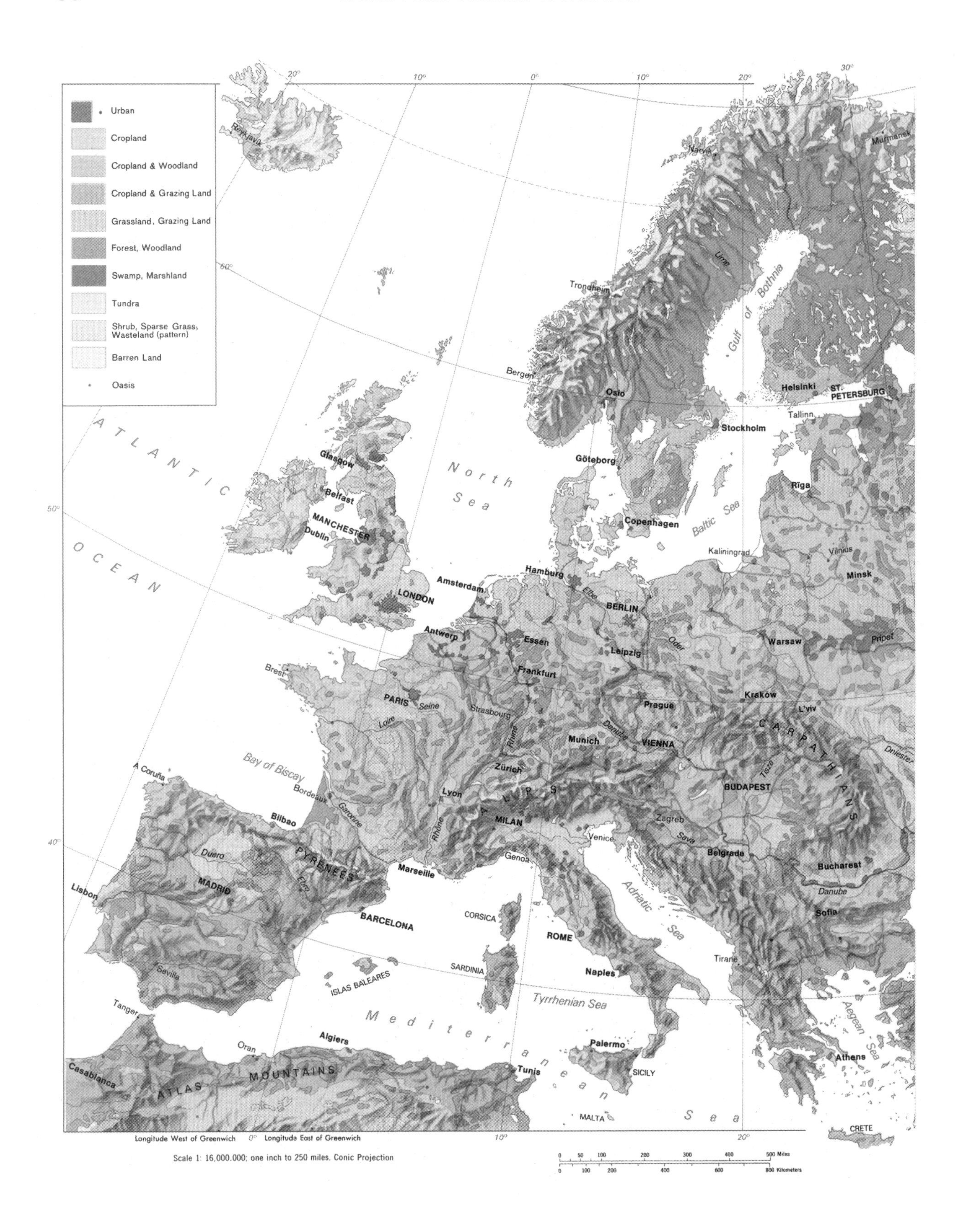
Urban
Cropland
Cropland & Woodland
Cropland & Grazing Land
Grassland, Grazing Land
Forest, Woodland
Swamp, Marshland
Tundra
Shrub, Sparse Grass, Wasteland (pattern)
Barren Land
Oasis
ATLANTIC OCEAN
North Sea
Baltic Sea
Gulf of Bothnia
Bay of Biscay
Mediterranean Sea
Tyrrhenian Sea
Adriatic Sea
Aegean Sea
ALPS
PYRENEES
CARPATHIANS
ATLAS MOUNTAINS
Reykjavík
Narvik
Murmansk
Trondheim
Bergen
Oslo
Stockholm
Helsinki
ST. PETERSBURG
Tallinn
Göteborg
Rīga
Copenhagen
Kaliningrad
Vilnius
Minsk
Glasgow
Belfast
MANCHESTER
Dublin
LONDON
Amsterdam
Hamburg
BERLIN
Antwerp
Essen
Leipzig
Warsaw
Frankfurt
Prague
Kraków
L'viv
Brest
PARIS
Seine
Loire
Strasbourg
Rhine
Munich
Danube
VIENNA
Zürich
Lyon
Rhône
Garonne
Bordeaux
A Coruña
Bilbao
Duero
MADRID
Ebro
Lisbon
Sevilla
Tanger
Casablanca
Oran
Algiers
Tunis
Marseille
BARCELONA
ISLAS BALEARES
CORSICA
SARDINIA
MILAN
Genoa
Venice
Zagreb
Sava
Belgrade
BUDAPEST
Tisza
Dniester
Bucharest
Sofia
ROME
Naples
Palermo
SICILY
MALTA
Tirana
Athens
CRETE
Elbe
Oder
Ume
Pripet
20°
10°
0°
30°
60°
50°
40°
Longitude West of Greenwich
Longitude East of Greenwich
Scale 1: 16,000,000; one inch to 250 miles. Conic Projection
Miles
Kilometers

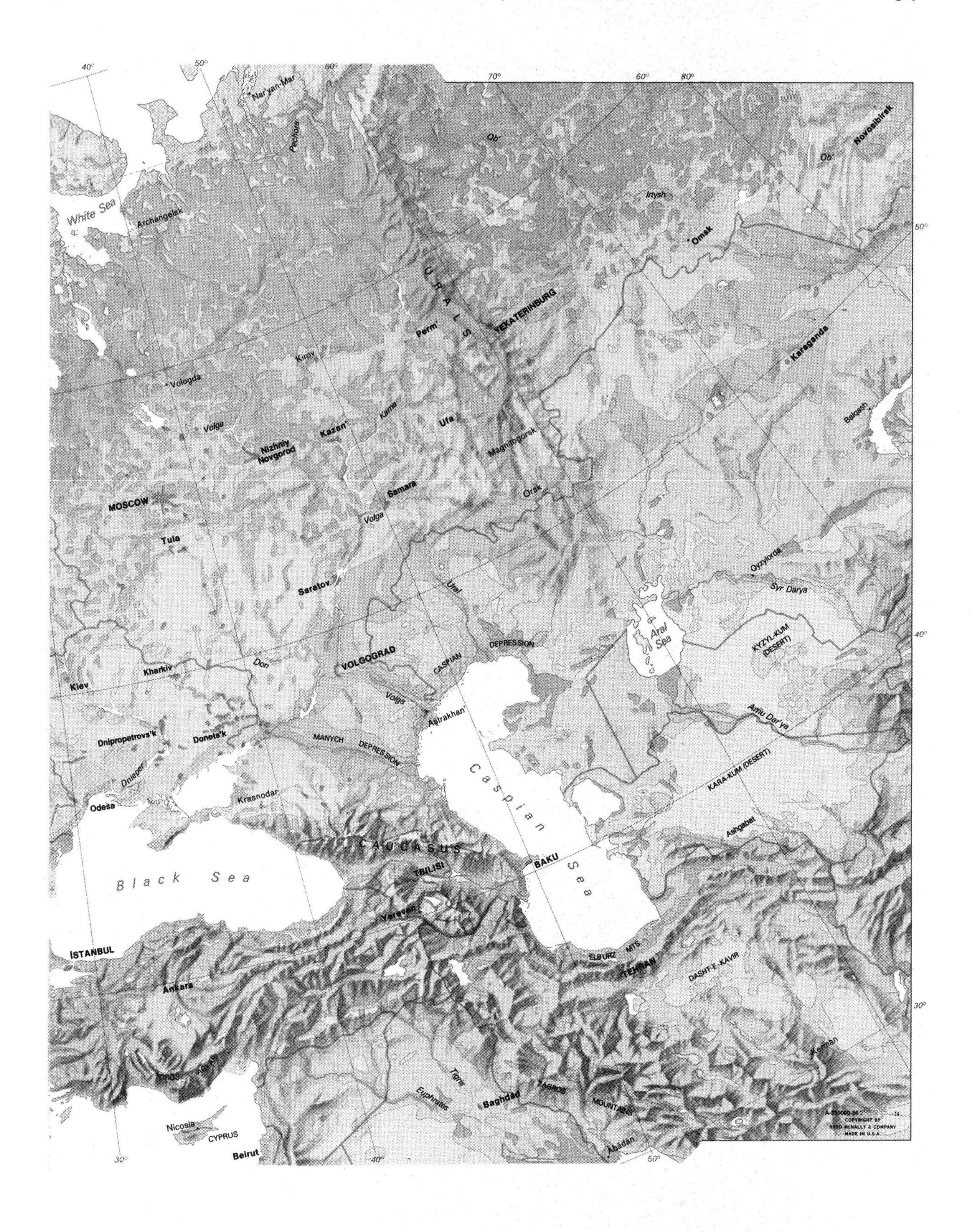
White Sea
Archangelsk
Nar'yan-Mar
Pechora
Ob'
Novosibirsk
Irtysh
Omsk
URALS
YEKATERINBURG
Perm'
Kirov
Karaganda
Vologda
Volga
Kama
Ufa
Balqash
Nizhniy Novgorod
Kazan'
Magnitogorsk
MOSCOW
Samara
Orsk
Tula
Saratov
Ural
Qyzylorda
Syr Darya
Aral Sea
KYZYL-KUM (DESERT)
DEPRESSION
CASPIAN
VOLGOGRAD
Don
Kharkiv
Kiev
Astrakhan'
Amu Dar'ya
Dnipropetrovs'k
Donets'k
MANYCH DEPRESSION
KARA-KUM (DESERT)
Dnieper
Caspian Sea
Krasnodar
Odesa
Ashgabat
CAUCASUS
BAKU
TBILISI
Black Sea
Yerevan
ISTANBUL
ELBURZ MTS.
TEHRAN
DASHT-E-KAVIR
Ankara
Kerman
TOROS
Tigris
Euphrates
Baghdad
ZAGROS MOUNTAINS
Nicosia
CYPRUS
Beirut
Abadan
COPYRIGHT BY RAND MCNALLY & COMPANY MADE IN U.S.A.

Alesund, Norway *Corbis Digital Stock*

## *Europe Physical Map*

Name: ______________________ Date: ______________________

See pages **148-149** in Goode's World Atlas (21e) for a full color map.

What are the major **mountain ranges** on the continent?

______________________

______________________

______________________

What are the major **rivers** on the continent?

______________________

______________________

______________________

What are the major **bodies of water** that can be found around the edges of continent?

______________________

______________________

______________________

List the country or countries where you find each of the following **features**.

Rhône River ______________________
Oder River ______________________
Seine River ______________________
Loire River ______________________
Sava River ______________________
Danube River ______________________
Carpathian Range ______________________
Alps ______________________
Elbe River ______________________

Where are the largest areas of forest in Europe?

Where are the major areas of cropland in Europe?

Are there connections between elevation and vegetation in Europe?

What countries have the largest expanses of "barren land"? What factors contribute to this pattern?

St Basil's Church, Moscow, Russia *Corbis Digital Stock*

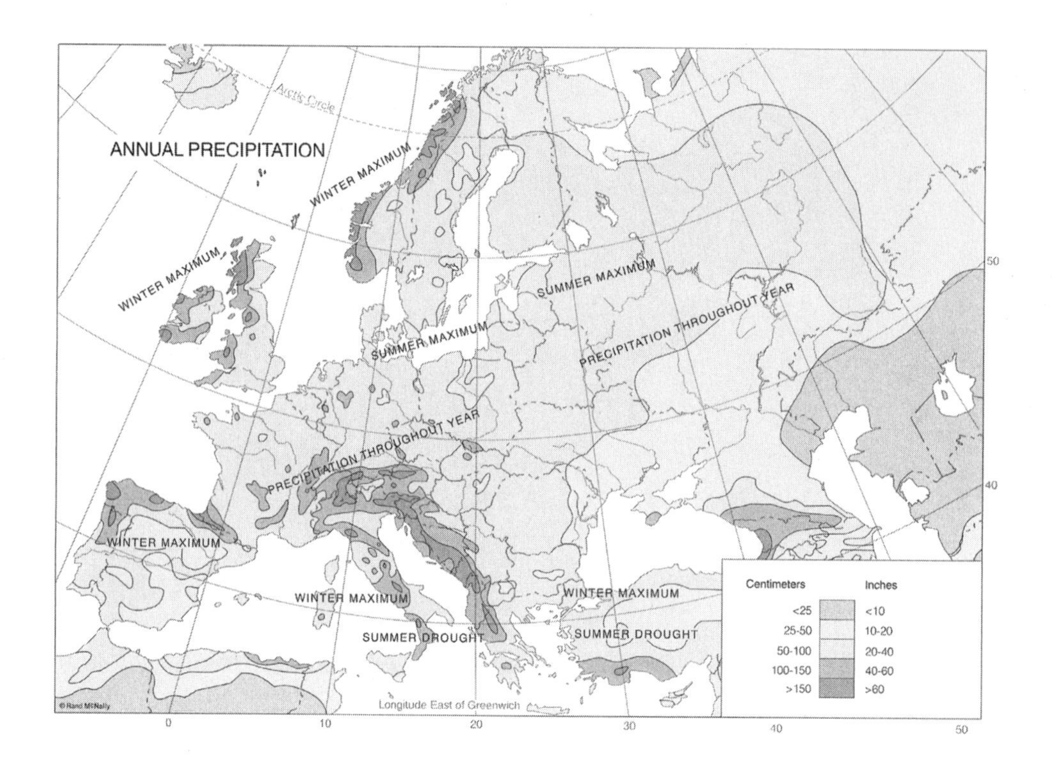


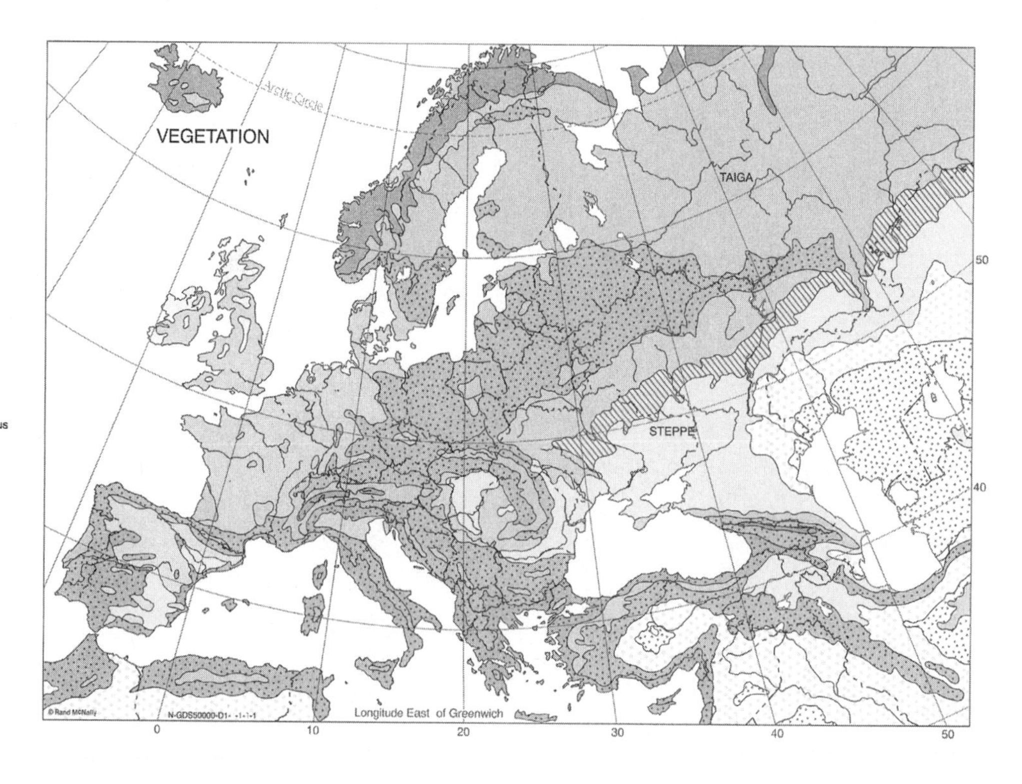


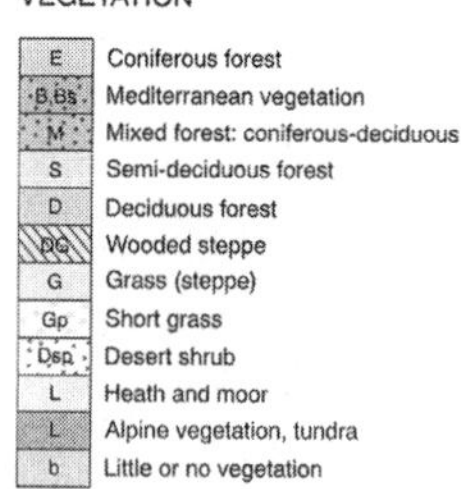


For explanation of letters in boxes, see Natural Vegetation Map by A. W. Kuchler, p. 24

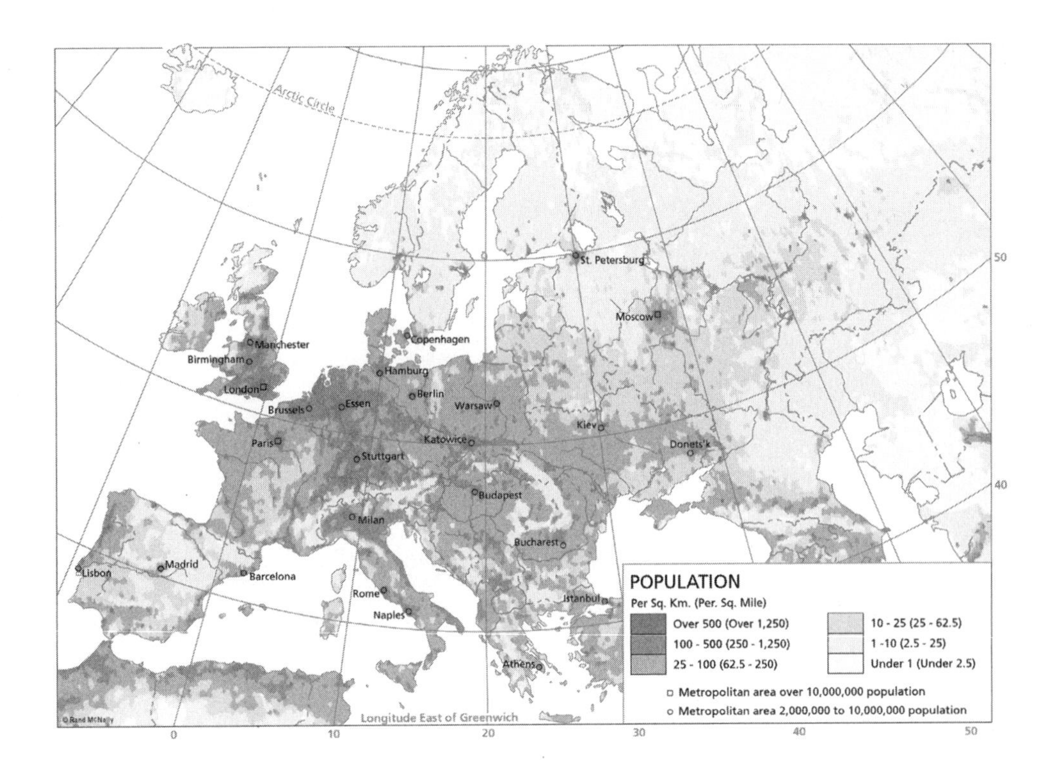
Arctic Circle
St. Petersburg
Moscow
Manchester
Birmingham
London
Copenhagen
Hamburg
Berlin
Brussels
Essen
Warsaw
Kiev
Paris
Katowice
Donets'k
Stuttgart
Budapest
Milan
Bucharest
Lisbon
Madrid
Barcelona
Rome
Naples
Istanbul
Athens
Longitude East of Greenwich
0
10
20
30
40
50
50
40
POPULATION
Per Sq. Km. (Per. Sq. Mile)
Over 500 (Over 1,250)
100 - 500 (250 - 1,250)
25 - 100 (62.5 - 250)
10 - 25 (25 - 62.5)
1 -10 (2.5 - 25)
Under 1 (Under 2.5)
Metropolitan area over 10,000,000 population
Metropolitan area 2,000,000 to 10,000,000 population

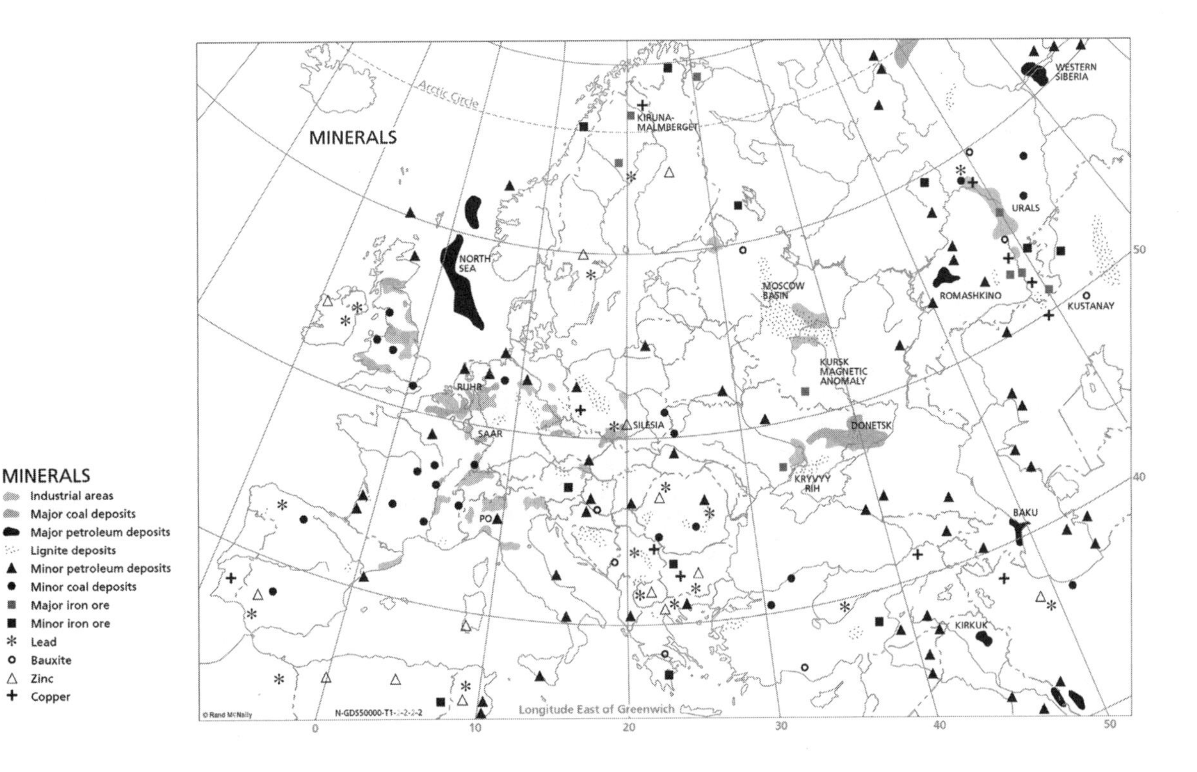
MINERALS
Arctic Circle
KIRUNA-MALMBERGET
WESTERN SIBERIA
URALS
NORTH SEA
MOSCOW BASIN
ROMASHKINO
KUSTANAY
KURSK MAGNETIC ANOMALY
RUHR
SILESIA
SAAR
DONETSK
KRYVYY RIH
PO
BAKU
KIRKUK
Longitude East of Greenwich
0
10
20
30
40
50
50
40
MINERALS
Industrial areas
Major coal deposits
Major petroleum deposits
Lignite deposits
Minor petroleum deposits
Minor coal deposits
Major iron ore
Minor iron ore
Lead
Bauxite
Zinc
Copper

Prague, Czech Republic *PhotoDisc, Inc.*

## *Europe Thematic Maps*

Name: ______________________________ Date: ______________________________

See pages **146-147** in Goode's World Atlas (21e) for full color maps.

What areas of Europe are most heavily populated?

What areas of Europe are least populated? Is there a connection between these areas and Europe's physical geography?

Where in Europe do you find areas of tundra?

What areas of Europe are largely covered in coniferous forests?

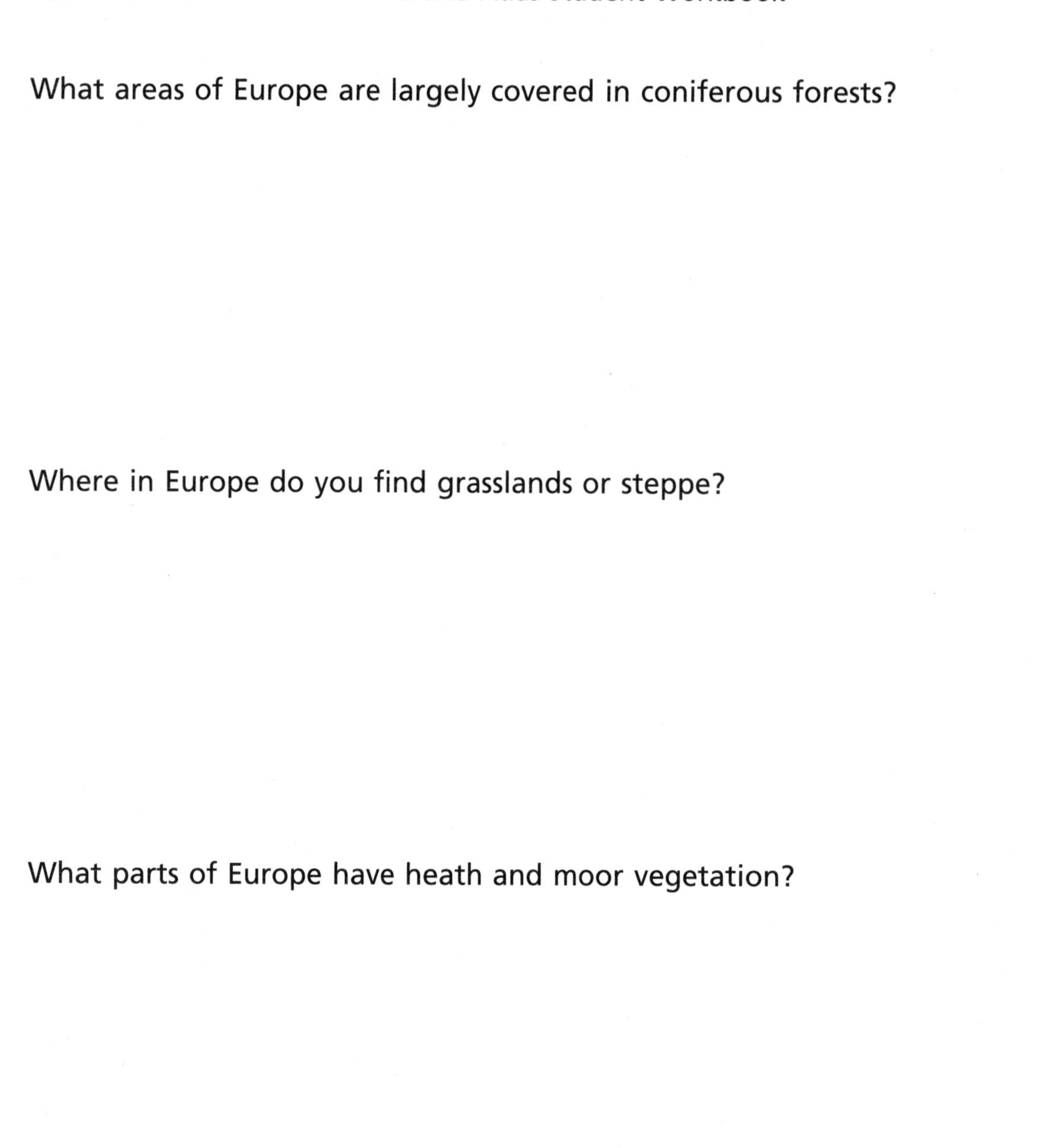

Where in Europe do you find grasslands or steppe?

What parts of Europe have heath and moor vegetation?

Examine the map of minerals. What factors might explain the location of Europe's major "industrial areas?"

L'Arc de Triomphe, Paris *Corbis Digital Stock*

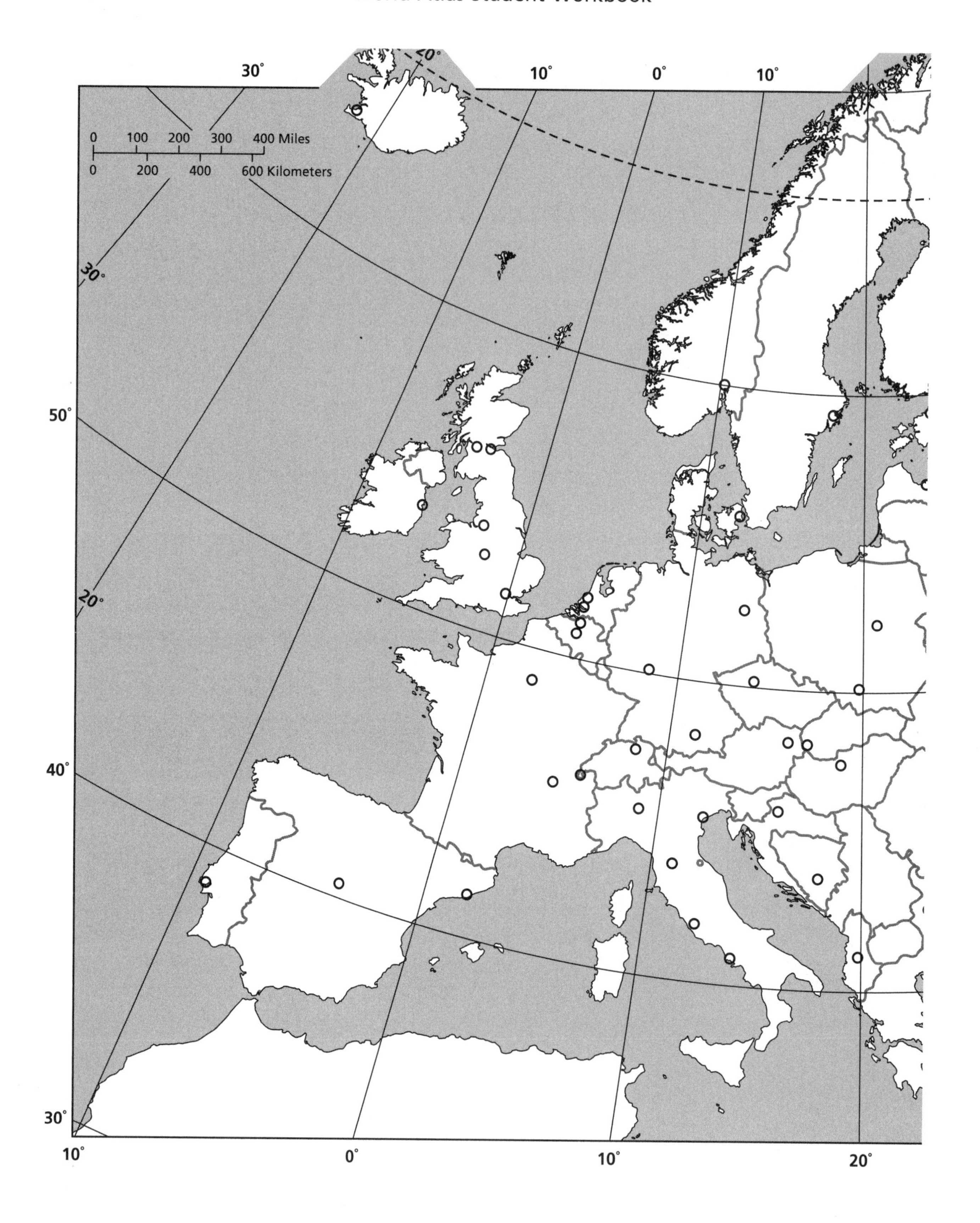
30°
20°
10°
0°
10°
0 100 200 300 400 Miles
0 200 400 600 Kilometers
30°
20°
50°
40°
30°
10°
0°
10°
20°

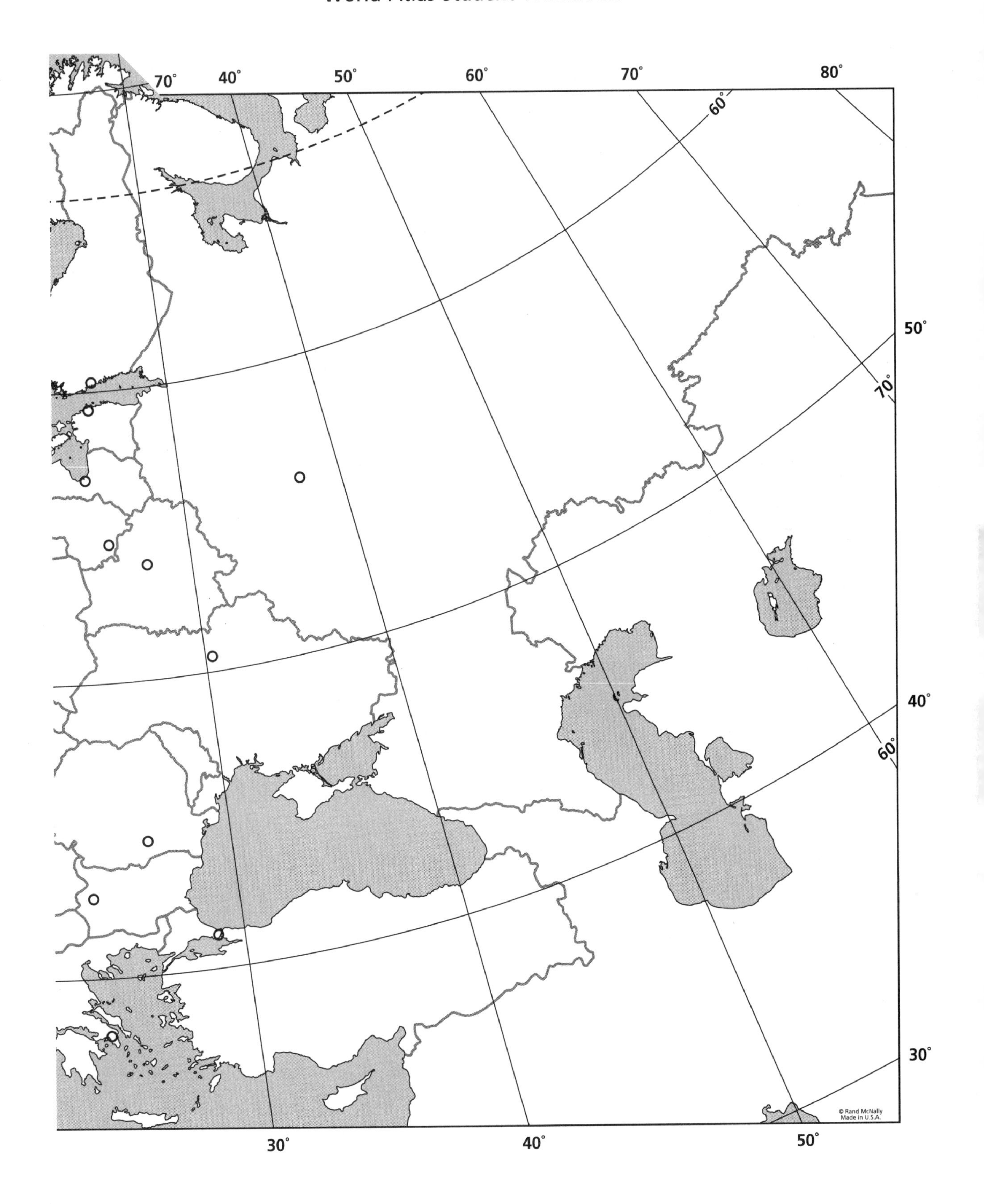

70°
40°
50°
60°
70°
80°
60°
50°
70°
40°
60°
30°
30°
40°
50°
© Rand McNally
Made in U.S.A.

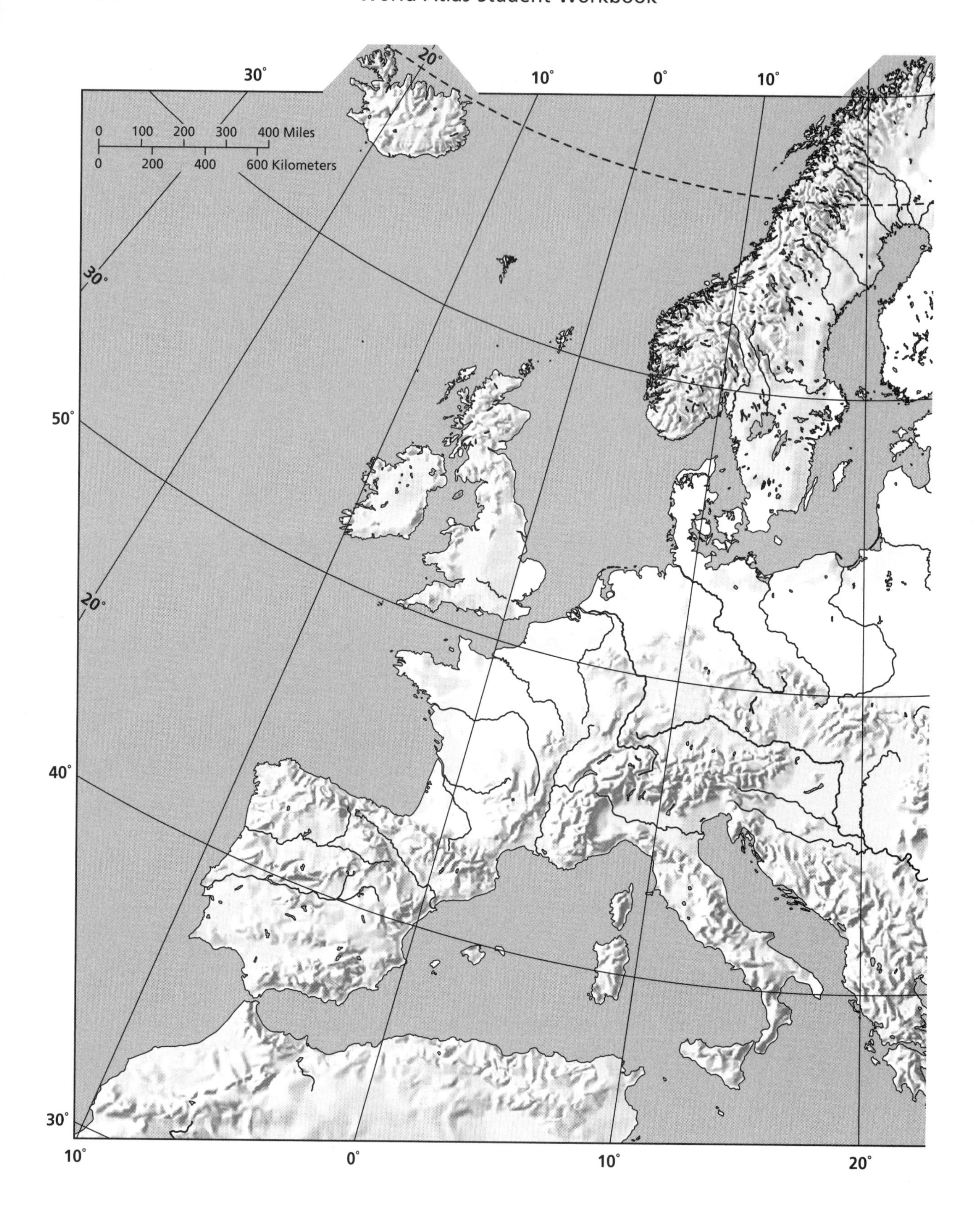

30°
20°
10°
0°
10°
0 100 200 300 400 Miles
0 200 400 600 Kilometers
30°
50°
20°
40°
30°
10°
0°
10°
20°

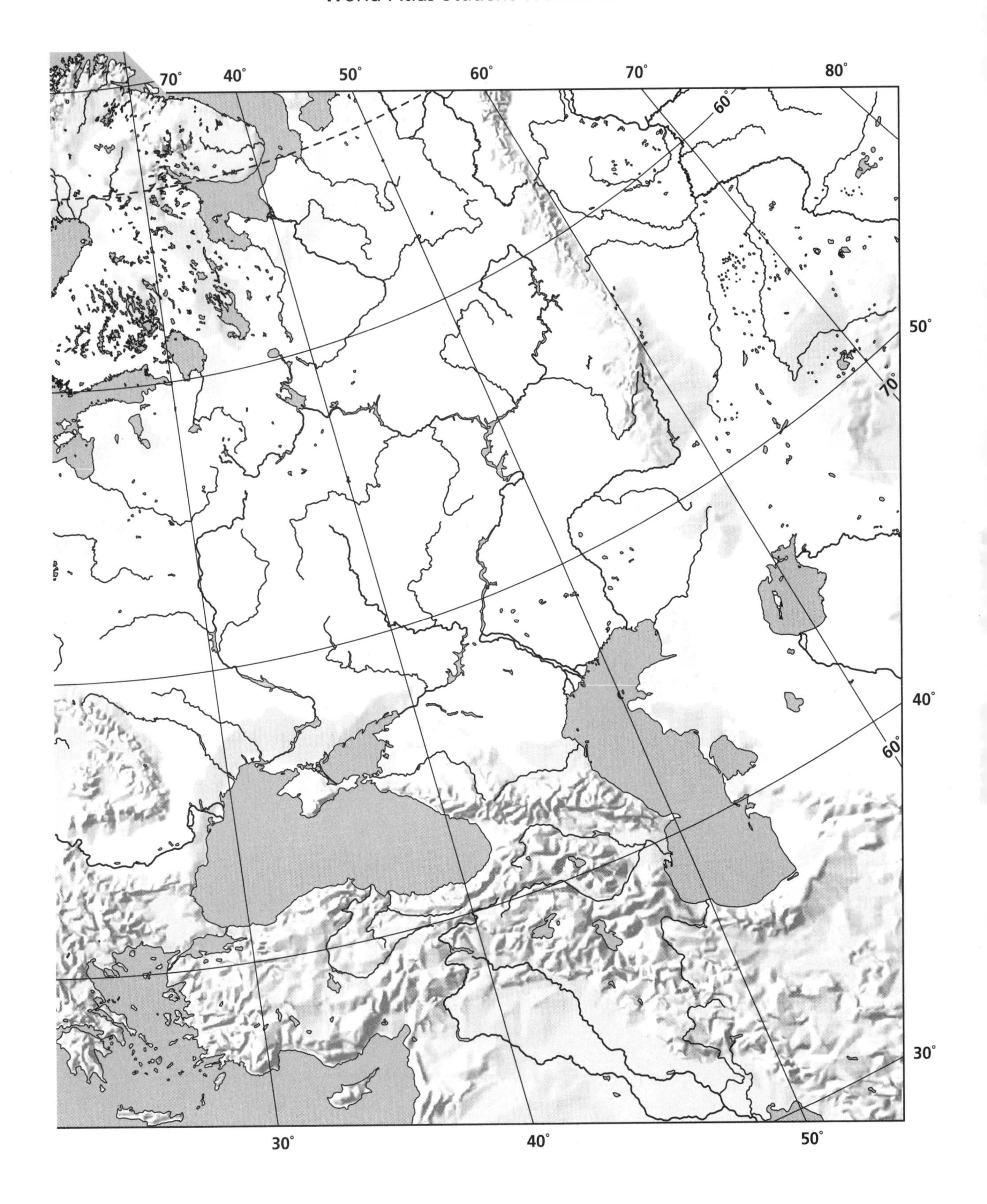
70°
40°
50°
60°
70°
80°
60°
50°
70°
40°
60°
30°
30°
40°
50°

Tower Bridge, London, United Kingdom *PhotoDisc, Inc.*

## Europe Blank Maps (Political / Physical)

Name: ______________________ Date: ______________________

See pages **148-149** & **154-155** in Goode's World Atlas (21e) for full color maps.

### Political Map

Label the following **countries** on the blank political map of Europe.

| | | |
|---|---|---|
| Albania | Greece | Poland |
| Andorra | Hungary | Portugal |
| Austria | Iceland | Romania |
| Belarus | Ireland | Russia |
| Belgium | Italy | San Marino |
| Bosnia and Herzegovina | Latvia | Serbia |
| Bulgaria | Liechtenstein | Slovakia |
| Croatia | Lithuania | Slovenia |
| Czech Republic | Luxembourg | Spain |
| Denmark | Macedonia | Sweden |
| Estonia | Malta | Switzerland |
| Finland | Moldova | Turkey |
| France | Netherlands | Ukraine |
| Germany | Norway | United Kingdom |

Label the following **cities** on the blank political map of Europe.

| | | |
|---|---|---|
| Antwerp | Helsinki | Prague |
| Athens | Istanbul | Reykjavik |
| Barcelona | Kiev | Riga |
| Berlin | Krakow | Rome |
| Birmingham | Lisbon | Rotterdam |
| Bratislava | London | Sarajevo |
| Brussels | Lyon | Sofia |
| Bucharest | Madrid | Stockholm |
| Budapest | Manchester | Tallinn |
| Copenhagen | Milan | Tirane |
| Dublin | Minsk | Venice |
| Edinburgh | Moscow | Vienna |
| Florence | Munich | Vilnius |
| Frankfurt | Naples | Warsaw |
| Geneva | Oslo | Zaghreb |
| Glasgow | Paris | Zurich |
| The Hague | | |

## Physical Map

Label the following **features** on the blank physical map of Europe.

### *Mountain Ranges*

| | |
|---|---|
| Alps<br>Carpathians | Pyrenees |

### *Rivers*

| | |
|---|---|
| Danube<br>Elbe<br>Loire<br>Oder | Rhine<br>Rhone<br>Sava<br>Seine |

### *Bodies of Water*

| | |
|---|---|
| Adriatic Sea<br>Aegean Sea<br>Atlantic Ocean<br>Baltic Sea<br>Bay of Biscay | Black Sea<br>Gulf of Bothnia<br>Mediterranean Sea<br>North Sea<br>Tyrrhenian Sea |

Bali, Indonesia *Corbis Digital Stock*

Scale 1:40 000 000; one inch to 630 miles. Lambert's Azimuthal, Equal Area Projection
Elevations and depressions are given in feet

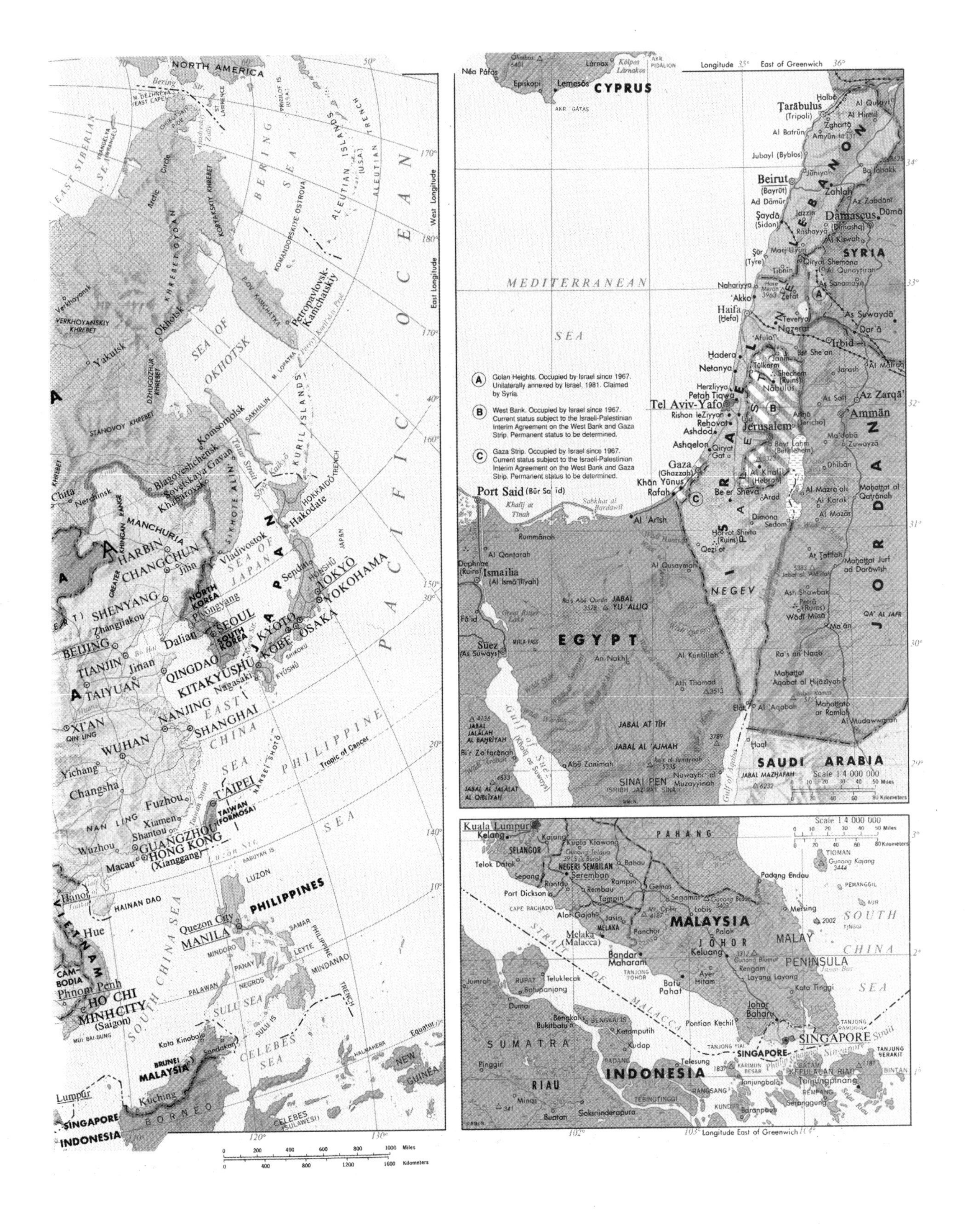
NORTH AMERICA
Bering Str.
SEA OF OKHOTSK
BERING SEA
ALEUTIAN ISLANDS
PACIFIC OCEAN
KURIL ISLANDS
Petropavlovsk-Kamchatskiy
Okhotsk
Yakutsk
Komsomolsk
Khabarovsk
Vladivostok
SEA OF JAPAN
HARBIN
CHANGCHUN
Jilin
MANCHURIA
SHENYANG
NORTH KOREA
P'yongyang
SEOUL
SOUTH KOREA
Dalian
BEIJING
TIANJIN
Jinan
QINGDAO
TAIYUAN
XI'AN
WUHAN
NANJING
SHANGHAI
EAST CHINA SEA
Chengdu
Yichang
Changsha
Fuzhou
Xiamen
Shantou
Wuzhou
GUANGZHOU
HONG KONG (Xianggang)
Macau
T'AIPEI
TAIWAN (FORMOSA)
KITAKYUSHU
Nagasaki
KYOTO
KOBE
OSAKA
TOKYO
YOKOHAMA
Sendai
Hakodate
JAPAN
HOKKAIDO
HONSHU
SHIKOKU
KYUSHU
PHILIPPINE SEA
Tropic of Cancer
Hanoi
HAINAN DAO
VIETNAM
Hue
CAMBODIA
Phnom Penh
HO CHI MINH CITY (Saigon)
SOUTH CHINA SEA
LUZON
PHILIPPINES
Quezon City
MANILA
MINDORO
PANAY
NEGROS
SAMAR
LEYTE
MINDANAO
PALAWAN
SULU SEA
CELEBES SEA
Kota Kinabalu
Sandakan
BRUNEI
MALAYSIA
Kuching
BORNEO
SINGAPORE
INDONESIA
Lumpur
CELEBES (SULAWESI)
HALMAHERA
NEW GUINEA
Equator
0 200 400 600 800 1000 Miles
0 400 800 1200 1600 Kilometers
CYPRUS
Lemesós
Lárnax
Néa Páfos
MEDITERRANEAN SEA
Longitude 35° East of Greenwich 36°
Tarābulus (Tripoli)
Beirut (Bayrūt)
Şaydā (Sidon)
Şūr (Tyre)
LEBANON
SYRIA
Damascus (Dimashq)
Haifa (Hefa)
Nazerat
Hadera
Netanya
Tel Aviv-Yafo
Ashdod
Ashqelon
Gaza (Ghazzah)
Khān Yūnus
Rafah
Jerusalem
Be'er Sheva'
ISRAEL
NEGEV
JORDAN
'Ammān
Az Zarqā'
Irbid
Port Said (Būr Sa'īd)
Ismailia (Al Ismā'īlīyah)
Suez (As Suways)
EGYPT
SINAI PEN.
Gulf of Suez
Gulf of Aqaba
Elat
Al 'Aqabah
SAUDI ARABIA
Ⓐ Golan Heights. Occupied by Israel since 1967. Unilaterally annexed by Israel, 1981. Claimed by Syria.
Ⓑ West Bank. Occupied by Israel since 1967. Current status subject to the Israeli-Palestinian Interim Agreement on the West Bank and Gaza Strip. Permanent status to be determined.
Ⓒ Gaza Strip. Occupied by Israel since 1967. Current status subject to the Israeli-Palestinian Interim Agreement on the West Bank and Gaza Strip. Permanent status to be determined.
Scale 1:4 000 000
Kuala Lumpur
Kelang
SELANGOR
PAHANG
NEGERI SEMBILAN
Seremban
Port Dickson
Melaka (Malacca)
MALAYSIA
JOHOR
Johor Baharu
MALAY PENINSULA
SOUTH CHINA SEA
STRAIT OF MALACCA
SUMATRA
RIAU
INDONESIA
SINGAPORE
Singapore Strait
Longitude East of Greenwich

Bangkok, Thailand *Corbis Digital Stock*

## *Asia Political Map*

Name: ______________________ Date: ______________________

See pages **194-195** in Goode's World Atlas (21e) for full color maps.

List a **country** or major **city** for each of the following letters:

| | |
|---|---|
| A ______________ | N ______________ |
| B ______________ | O ______________ |
| C ______________ | P ______________ |
| D ______________ | Q ______________ |
| E ______________ | R ______________ |
| F ______________ | S ______________ |
| G ______________ | T ______________ |
| H ______________ | U ______________ |
| I ______________ | V ______________ |
| J ______________ | W ______________ |
| K ______________ | X ______________ |
| L ______________ | Y ______________ |
| M ______________ | Z ______________ |

List all the countries that border China.

List all the countries that border Russia.

Which countries in Asia are landlocked?

Which countries are largely composed of islands?

Which countries in Asia are south of the Equator?

Which countries border Israel?

Which countries border Iraq?

Which countries border the Caspian Sea?

List all the countries on the Arabian Peninsula.

Which countries border India?

Itsukushima Jinja Shrine, Miyajima, Japan *Corbis Digital Stock*

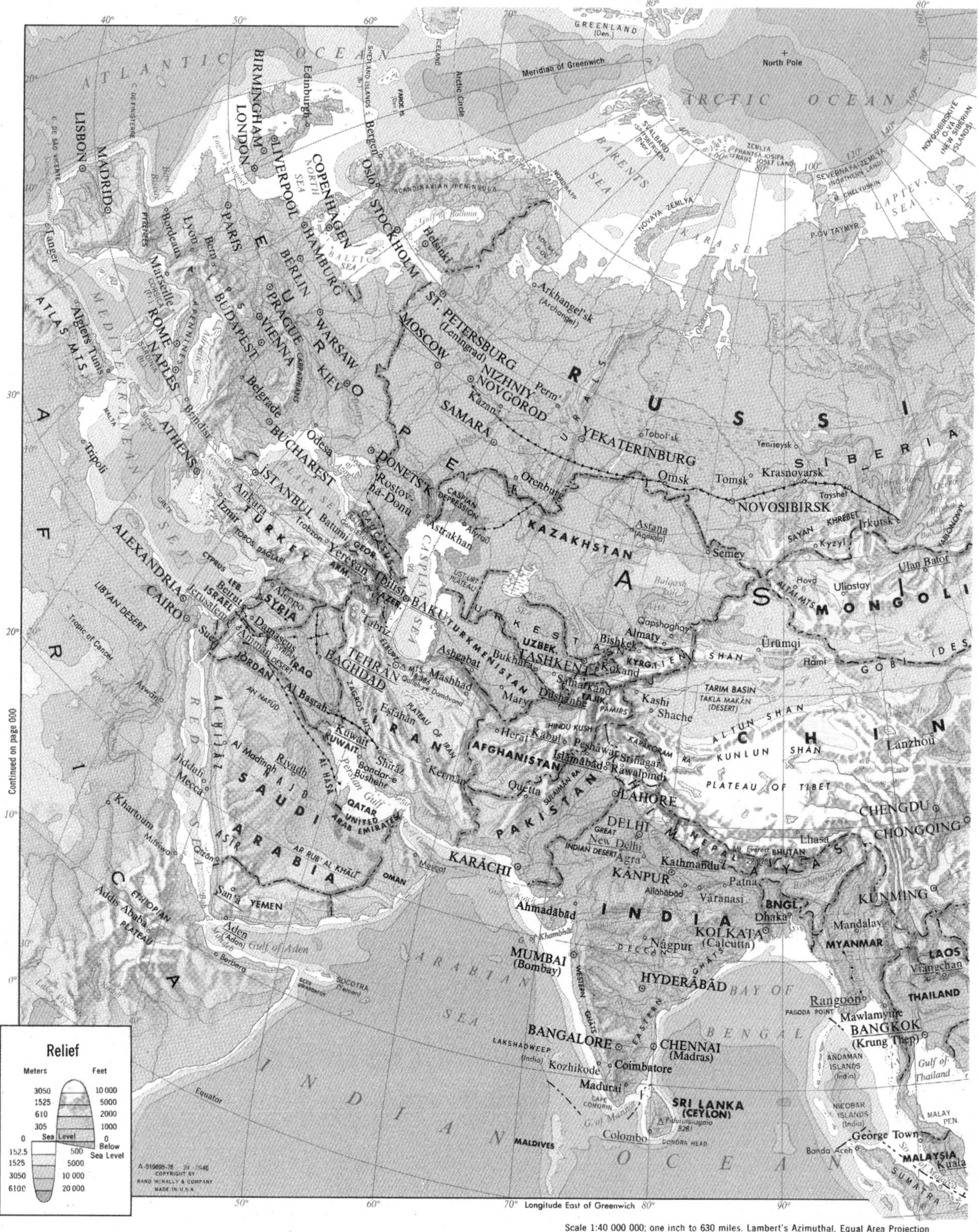


Continued on page 000

Scale 1:40 000 000; one inch to 630 miles. Lambert's Azimuthal, Equal Area Projection
Elevations and depressions are given in feet

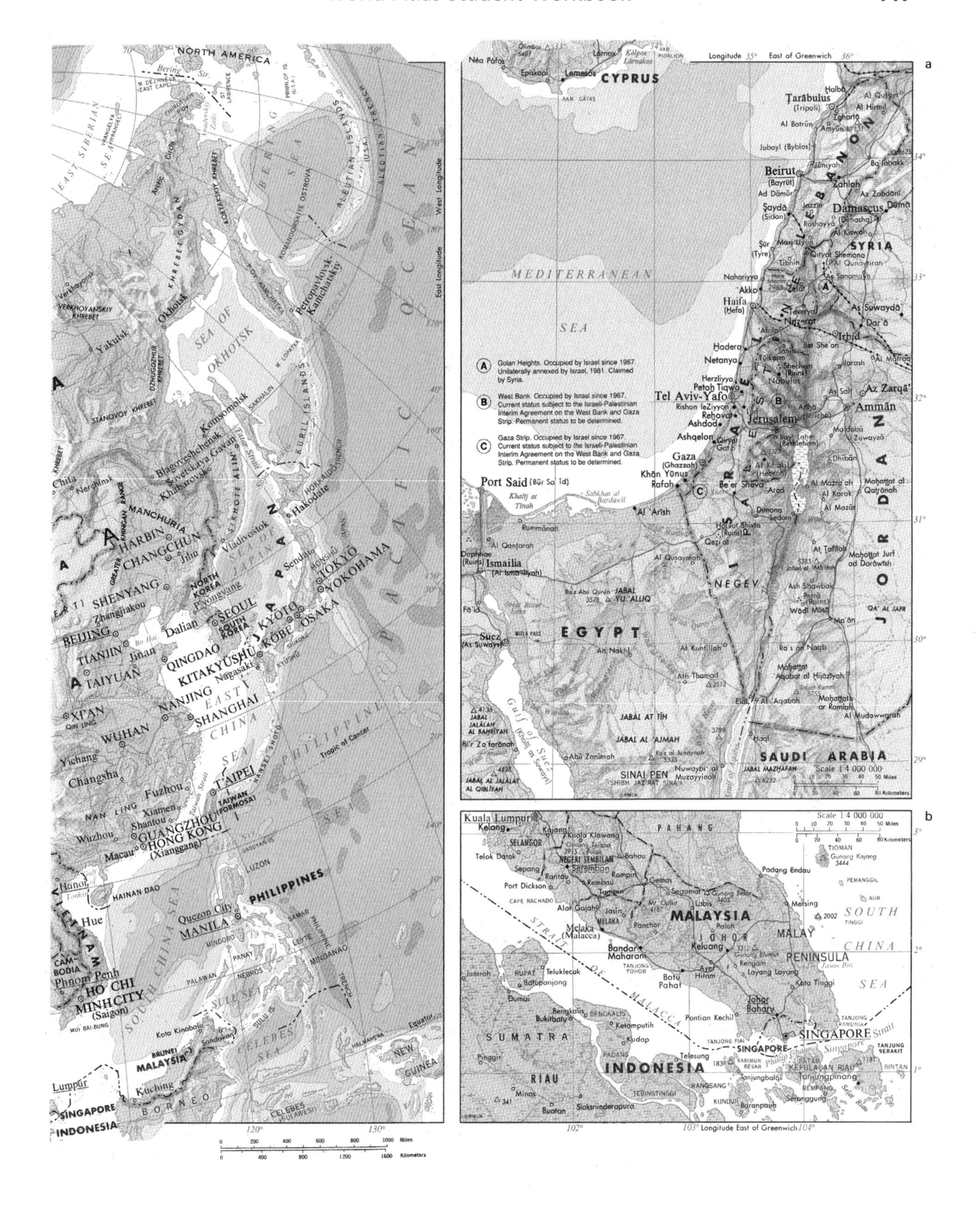
NORTH AMERICA
SEA OF OKHOTSK
SEA OF JAPAN
EAST CHINA SEA
SOUTH CHINA SEA
PHILIPPINE SEA
PACIFIC OCEAN
Tropic of Cancer
Equator
HARBIN
CHANGCHUN
SHENYANG
BEIJING
TIANJIN
TAIYUAN
QINGDAO
NANJING
SHANGHAI
WUHAN
XI'AN
GUANGZHOU
HONG KONG (Xianggang)
TAIPEI
SEOUL
KITAKYUSHU
KOBE
OSAKA
KYOTO
TOKYO
YOKOHAMA
MANILA
Quezon City
PHILIPPINES
HO CHI MINH CITY (Saigon)
Phnom Penh
Hanoi
Hue
SINGAPORE
INDONESIA
MALAYSIA
BRUNEI
NEW GUINEA
BORNEO
120°
130°
0 200 400 600 800 1000 Miles
0 400 800 1200 1600 Kilometers
a
CYPRUS
Longitude 35° East of Greenwich 36°
MEDITERRANEAN SEA
LEBANON
SYRIA
ISRAEL
JORDAN
EGYPT
SAUDI ARABIA
NEGEV
Beirut
Damascus
Haifa
Tel Aviv-Yafo
Jerusalem
'Ammān
Gaza
Port Said
Ismailia
Suez
Elat
Al 'Aqabah
Gulf of Suez
Gulf of Aqaba
SINAI PEN.
A Golan Heights. Occupied by Israel since 1967. Unilaterally annexed by Israel, 1981. Claimed by Syria.
B West Bank. Occupied by Israel since 1967. Current status subject to the Israeli-Palestinian Interim Agreement on the West Bank and Gaza Strip. Permanent status to be determined.
C Gaza Strip. Occupied by Israel since 1967. Current status subject to the Israeli-Palestinian Interim Agreement on the West Bank and Gaza Strip. Permanent status to be determined.
Scale 1:4 000 000
b
Kuala Lumpur
MALAYSIA
SINGAPORE
INDONESIA
SUMATRA
RIAU
STRAIT OF MALACCA
SOUTH CHINA SEA
MALAY PENINSULA
PAHANG
JOHOR
Melaka (Malacca)
Scale 1:4 000 000
102°
103° Longitude East of Greenwich 104°

Phnom Penh, Cambodia *PhotoDisc, Inc.*

## *Asia Physical Map*

Name: ______________________ Date: ______________________

See pages **192-193** in Goode's World Atlas (21e) for a full color map.

What are the major **mountain ranges** on the continent?

______________________

______________________

______________________

What are the major **rivers** on the continent?

______________________

______________________

______________________

List the country or countries where you find each of the following **features**.

Gobi Desert ______________________
Dasht-e Kavir ______________________
Hindu Kush Range ______________________
Himalayas ______________________
Ar Rub' al Khali ______________________
Takla Makan Desert ______________________
Plateau of Tibet ______________________
Honshu Island ______________________
Island of Borneo ______________________
Urals Range ______________________
Aral Sea ______________________
Lake Baikal ______________________
Island of Sumatra ______________________
Huang (Yellow) River ______________________
Sakhalin Island ______________________
Ganges River ______________________
Indus River ______________________
Brahmaputra River ______________________
Tigris River ______________________
Euphrates River ______________________
Altai Mountains ______________________
Mekong River ______________________

Is there any pattern to the location of deserts in Asia?

What effects do the Himalayas have on the vegetation north and south of the mountains?

Does the physical geography of Asia help to explain why the Mongols were able to expand from modern-day Mongolia to Europe?

Where are large forests located on the continent?

What do areas of tundra have in common in Asia?

Where are large areas of cropland in Asia? Is there a pattern?

Taj Mahal, Agra, India *Corbis Digital Stock*

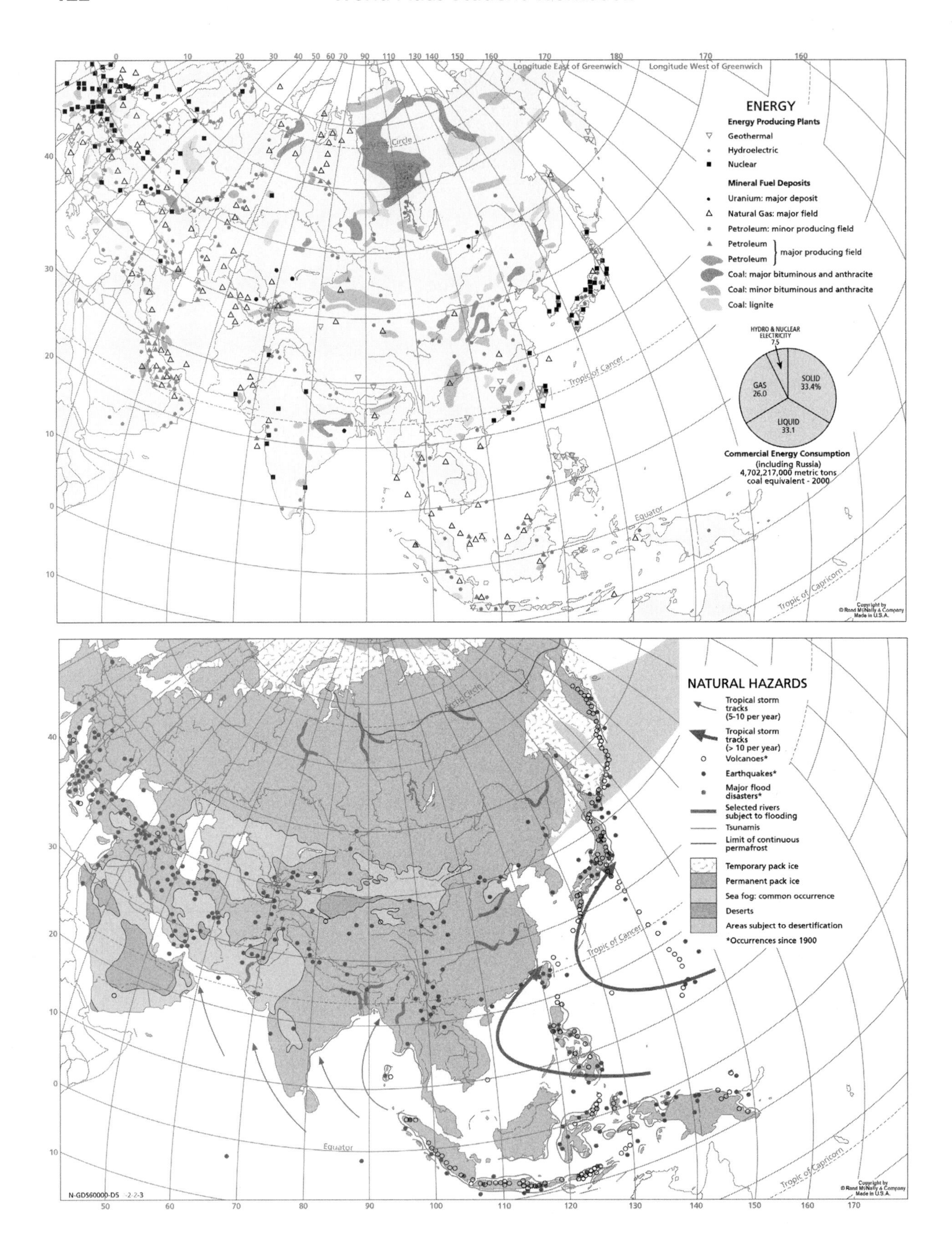
ENERGY
Energy Producing Plants
Geothermal
Hydroelectric
Nuclear
Mineral Fuel Deposits
Uranium: major deposit
Natural Gas: major field
Petroleum: minor producing field
Petroleum
Petroleum
major producing field
Coal: major bituminous and anthracite
Coal: minor bituminous and anthracite
Coal: lignite
HYDRO & NUCLEAR ELECTRICITY 7.5
GAS 26.0
SOLID 33.4%
LIQUID 33.1
Commercial Energy Consumption
(including Russia)
4,702,217,000 metric tons
coal equivalent - 2000
Longitude East of Greenwich
Longitude West of Greenwich
Arctic Circle
Tropic of Cancer
Equator
Tropic of Capricorn
NATURAL HAZARDS
Tropical storm tracks (5-10 per year)
Tropical storm tracks (> 10 per year)
Volcanoes*
Earthquakes*
Major flood disasters*
Selected rivers subject to flooding
Tsunamis
Limit of continuous permafrost
Temporary pack ice
Permanent pack ice
Sea fog: common occurrence
Deserts
Areas subject to desertification
*Occurrences since 1900
N-GDS60000-D5 -2-2-3

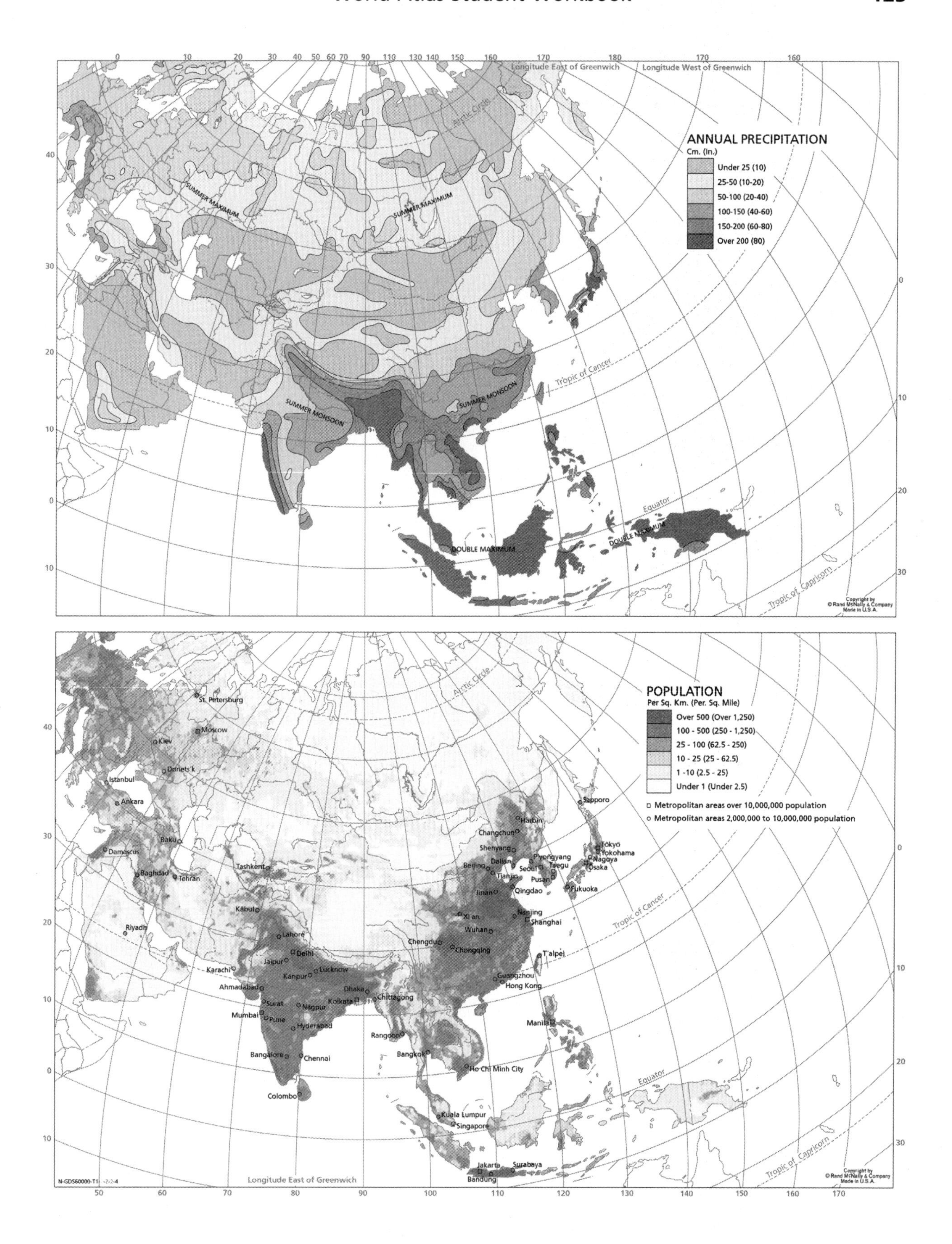
Longitude East of Greenwich
Longitude West of Greenwich
ANNUAL PRECIPITATION
Cm. (In.)
Under 25 (10)
25-50 (10-20)
50-100 (20-40)
100-150 (40-60)
150-200 (60-80)
Over 200 (80)
Arctic Circle
SUMMER MAXIMUM
SUMMER MAXIMUM
SUMMER MONSOON
SUMMER MONSOON
Tropic of Cancer
Equator
DOUBLE MAXIMUM
DOUBLE MAXIMUM
Tropic of Capricorn
Copyright by © Rand McNally & Company Made in U.S.A.
POPULATION
Per Sq. Km. (Per. Sq. Mile)
Over 500 (Over 1,250)
100 - 500 (250 - 1,250)
25 - 100 (62.5 - 250)
10 - 25 (25 - 62.5)
1 -10 (2.5 - 25)
Under 1 (Under 2.5)
Metropolitan areas over 10,000,000 population
Metropolitan areas 2,000,000 to 10,000,000 population
St. Petersburg
Moscow
Kiev
Donets'k
Istanbul
Ankara
Baku
Damascus
Baghdad
Tehran
Tashkent
Kabul
Riyadh
Karachi
Lahore
Delhi
Jaipur
Lucknow
Kanpur
Ahmadabad
Dhaka
Kolkata
Chittagong
Surat
Nagpur
Mumbai
Pune
Hyderabad
Bangalore
Chennai
Colombo
Rangoon
Bangkok
Ho Chi Minh City
Kuala Lumpur
Singapore
Jakarta
Surabaya
Bandung
Manila
Hong Kong
Guangzhou
T'aipei
Chengdu
Chongqing
Wuhan
Xi'an
Nanjing
Shanghai
Jinan
Qingdao
Tianjin
Beijing
Dalian
Shenyang
Changchun
Harbin
P'yongyang
Seoul
Taegu
Pusan
Sapporo
Tokyo
Yokohama
Nagoya
Osaka
Fukuoka
Arctic Circle
Tropic of Cancer
Equator
Tropic of Capricorn
Longitude East of Greenwich
Copyright by © Rand McNally & Company Made in U.S.A.

Beijing, China *Corbis Digital Stock*

## *Asia Thematic Maps*

Name: ______________________ Date: ______________________

See pages **189-190** in Goode's World Atlas (21e) for full color maps.

Where are the largest population clusters in Asia? What factors might explain these locations?

Why does the population density abruptly and dramatically decline to the north of India?

What areas of Asia are the least populated? Do these areas have anything in common?

What are the primary natural hazards that affect Asia?

Is there a pattern to the location of volcanoes and earthquakes in Asia?

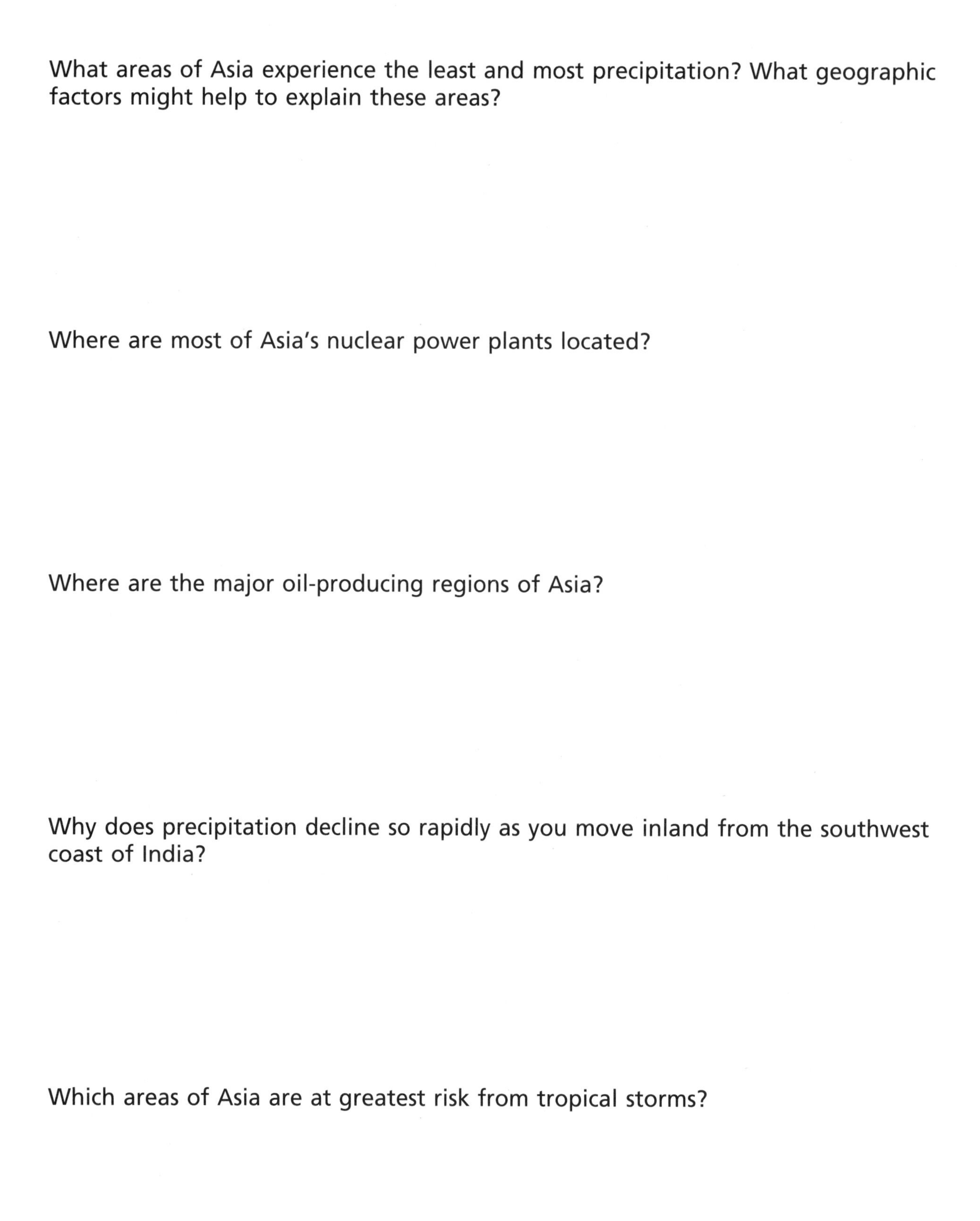

What areas of Asia experience the least and most precipitation? What geographic factors might help to explain these areas?

Where are most of Asia's nuclear power plants located?

Where are the major oil-producing regions of Asia?

Why does precipitation decline so rapidly as you move inland from the southwest coast of India?

Which areas of Asia are at greatest risk from tropical storms?

Hindu Temple, Singapore *Corbis Digital Stock*

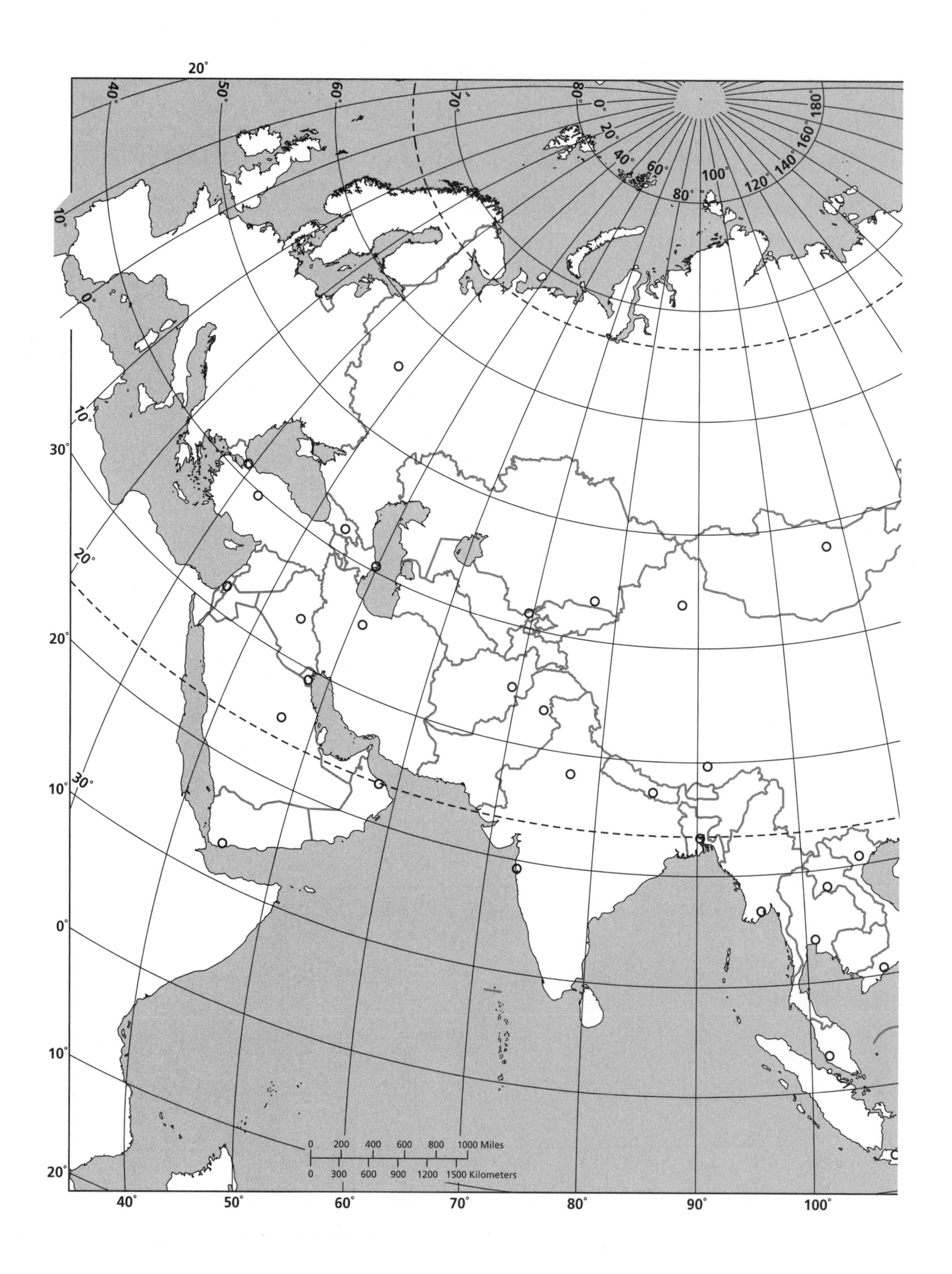
0 200 400 600 800 1000 Miles
0 300 600 900 1200 1500 Kilometers

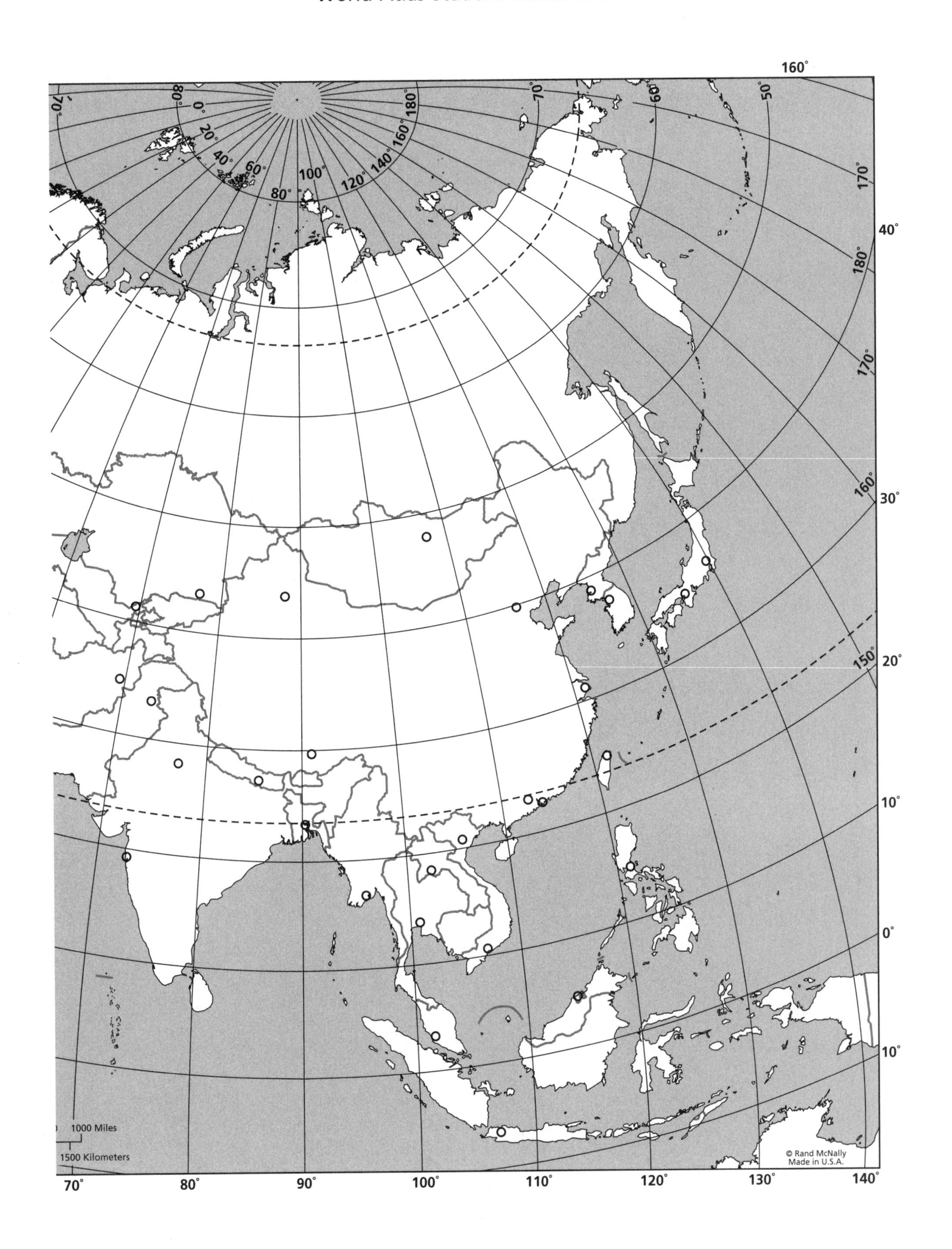
160°
170°
180°
170°
160°
150°
40°
30°
20°
10°
0°
10°
70°
80°
90°
100°
110°
120°
130°
140°
1000 Miles
1500 Kilometers
© Rand McNally
Made in U.S.A.

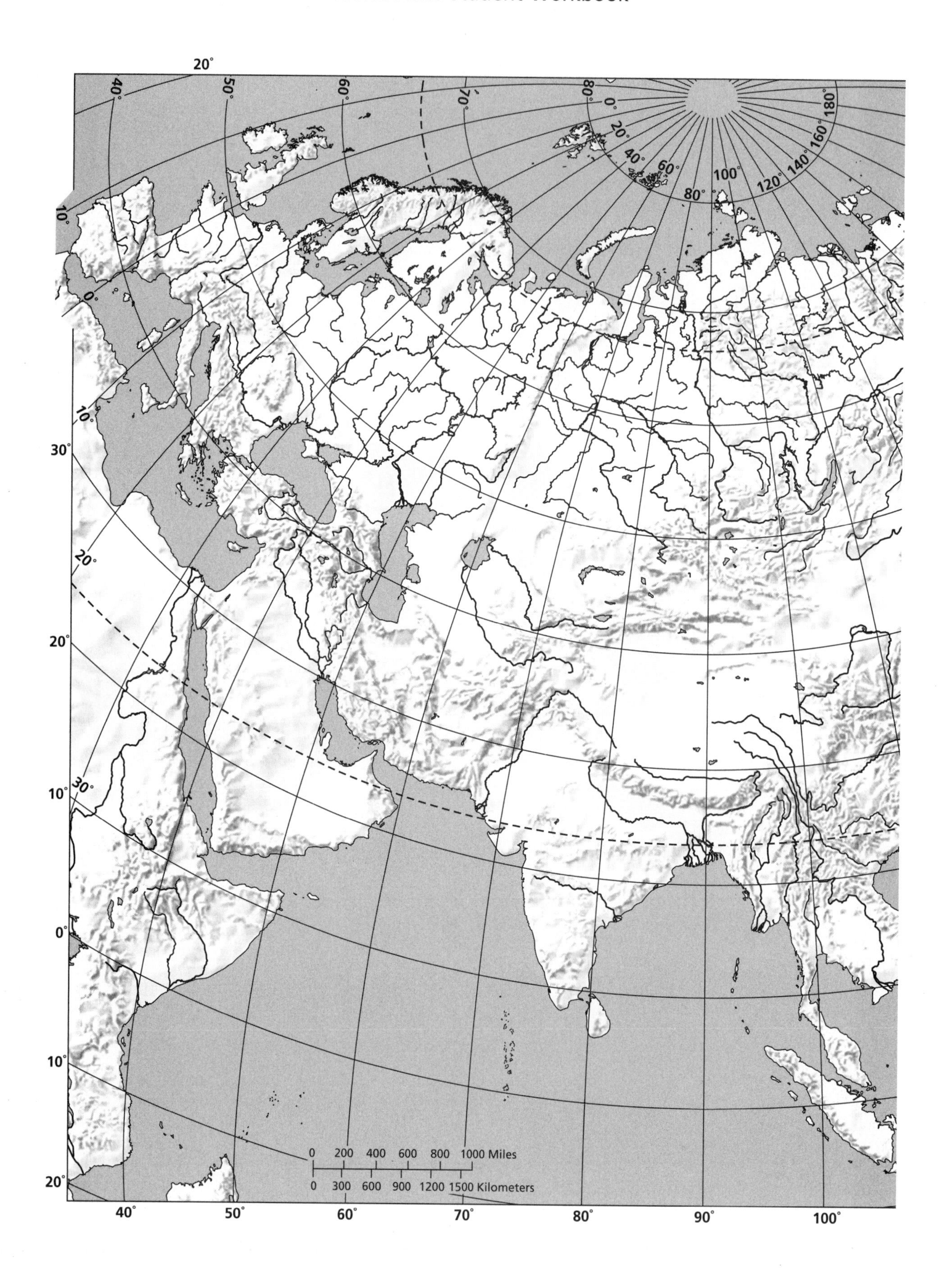

20°
40°
50°
60°
70°
80°
0°
20°
40°
60°
80°
100°
120°
140°
160°
180°
10°
0°
10°
20°
30°
30°
20°
10°
0°
10°
20°
0 200 400 600 800 1000 Miles
0 300 600 900 1200 1500 Kilometers
40°
50°
60°
70°
80°
90°
100°

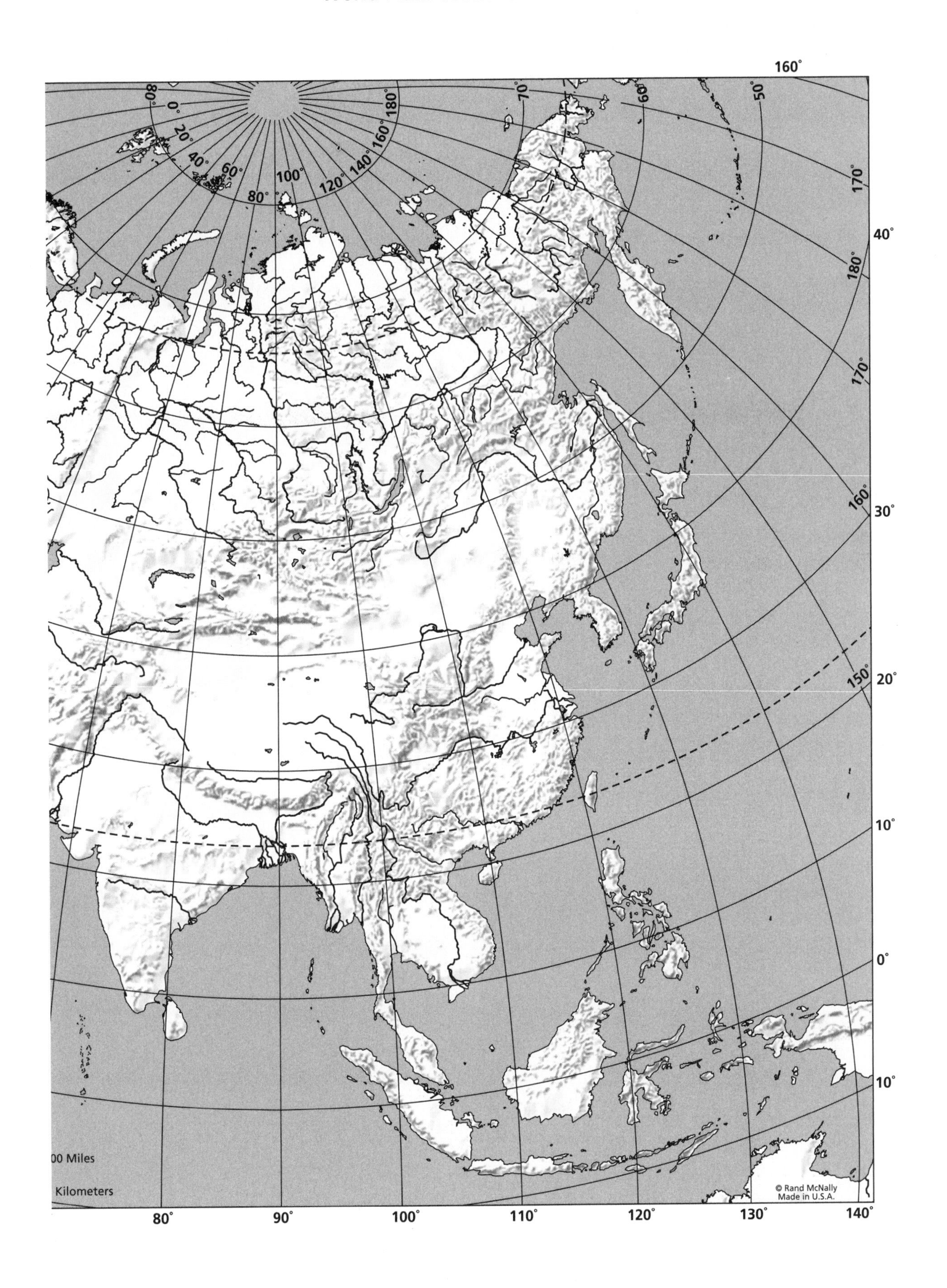
160°
80°
0°
20°
40°
60°
80°
100°
120°
140°
160°
180°
70°
60°
50°
170°
40°
180°
170°
160°
30°
150°
20°
10°
0°
10°
00 Miles
Kilometers
© Rand McNally
Made in U.S.A.
80°
90°
100°
110°
120°
130°
140°

Nepal *PhotoDisc, Inc.*

# *Asia Blank Maps (Political / Physical)*

Name: ______________________ Date: ______________________

See pages **192-193** in Goode's World Atlas (21e) for full color maps.

## Political Map

Label the following **countries** on the blank political map of Asia.

| | | |
|---|---|---|
| Afghanistan | Jordan | Russia |
| Armenia | Kazakhstan | Saudi Arabia |
| Azerbaijan | Kuwait | Singapore |
| Bahrain | Kyrgyzstan | South Korea |
| Bangladesh | Laos | Sri Lanka |
| Bhutan | Lebanon | Syria |
| Brunei | Malaysia | Taiwan |
| Cambodia | Mongolia | Tajikistan |
| China | Myanmar (Burma) | Thailand |
| Georgia | Nepal | Turkey |
| India | North Korea | Turkmenistan |
| Indonesia | Oman | United Arab Emirates |
| Iraq | Pakistan | Uzbekistan |
| Israel | Philippines | Vietnam |
| Japan | Qatar | Yemen |

Label the following **cities** on the blank political map of Asia.

| | | |
|---|---|---|
| Aden | Istanbul | P'yongyang |
| Almaty | Jakarta | Rangoon |
| Ankara | Jerusalem | Riyadh |
| Baghdad | Kabul | Seoul |
| Baku | Kathmandu | Shanghai |
| Bandar Seri Begawan | Kuala Lumpur | Taipei |
| Bangkok | Kuwait City | Tashkent |
| Beijing | Lhasa | Tbilisi |
| Dhaka | Manila | Tehran |
| Guangzhou | Moscow | Tokyo |
| Hanoi | Mumbai | Ulan Bator |
| Ho Chi Minh City | Muscat | Urumqi |
| Hong Kong | New Delhi | Viangchan |
| Islamabad | Osaka | |

## Physical Map

Label the following **features** on the blank physical map of Asia.

### *Mountain Ranges*

| | |
|---|---|
| Altai<br>Caucasus<br>Eastern Ghats<br>Himalayas<br>Hindu Kush | Tien Shan<br>Urals<br>Western Ghats<br>Zagros |

### *Rivers*

| | |
|---|---|
| Amur<br>Brahmaputra<br>Don<br>Euphrates<br>Ganges<br>Huang (Yellow)<br>Indus | Irtysh<br>Lena<br>Mekong<br>Ob<br>Tigris<br>Volga<br>Yangtze |

### *Bodies of Water*

| | |
|---|---|
| Andaman Sea<br>Arabian Sea<br>Aral Sea<br>Bay of Bengal<br>Black Sea<br>Caspian Sea<br>Celebes Sea<br>East China Sea<br>Indian Ocean | Java Sea<br>Lake Baikal<br>Pacific Ocean<br>Persian Gulf<br>Philippine Sea<br>Red Sea<br>Sea of Japan<br>Yellow Sea |

### *Islands*

| | |
|---|---|
| Borneo<br>Celebes<br>Java<br>Sumatra<br>Mindanao | Honshu<br>Kyushu<br>Hokkaido<br>Sakhalin<br>Hainan Dao |

Antarctica *Corbis Digital Stock*

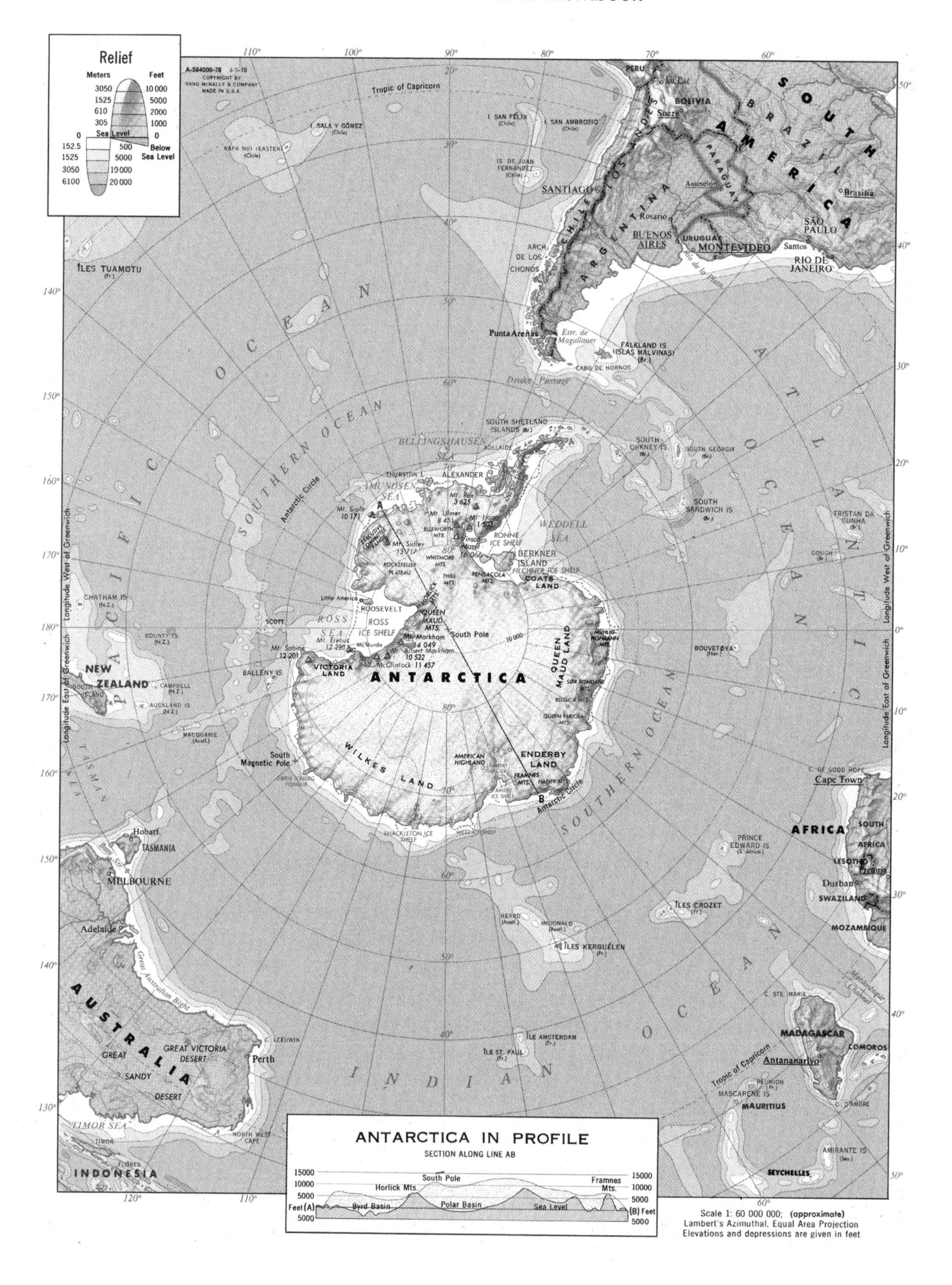


Scale 1: 60 000 000; (approximate)
Lambert's Azimuthal, Equal Area Projection
Elevations and depressions are given in feet

## *Antarctica Physical/Political Map*

Name: ______________________ Date: ______________________

See page **224** in Goode's World Atlas (21e) for a full color map.

What are the five highest **mountains** on the continent?

______________

______________

______________

______________

______________

What are the major **ice shelves** on the continent?

What are the **seas** that surround Antarctica?

What are the five **lands** that can be found on Antarctica?

What continent is closest to Antarctica?

Whale Bones, Antarctica *PhotoDisc, Inc.*

Antarctica *Corbis Digital Stock*

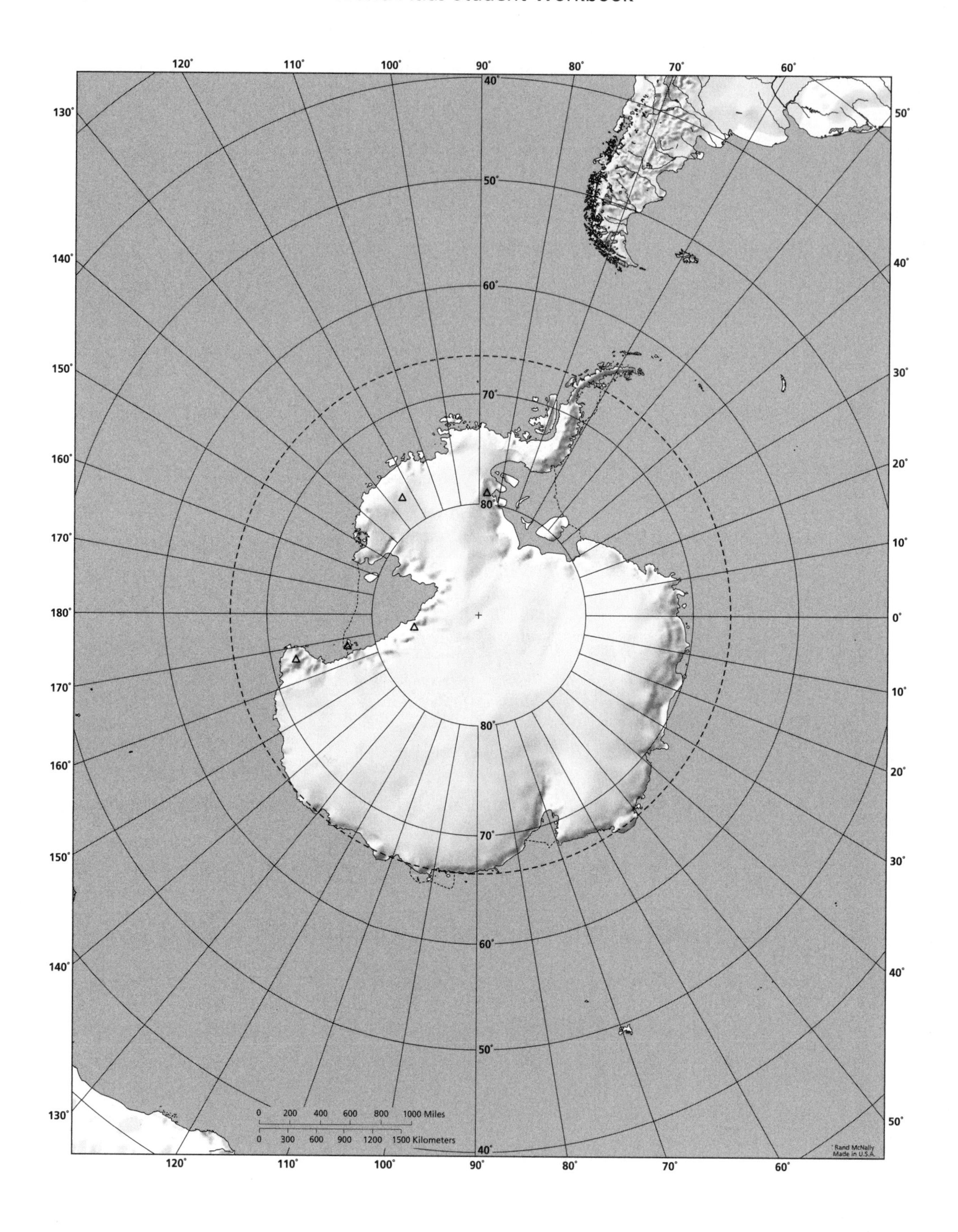
120°
110°
100°
90°
80°
70°
60°
130°
140°
150°
160°
170°
180°
170°
160°
150°
140°
130°
50°
40°
30°
20°
10°
0°
10°
20°
30°
40°
50°
40°
50°
60°
70°
80°
80°
70°
60°
50°
40°
0 200 400 600 800 1000 Miles
0 300 600 900 1200 1500 Kilometers
© Rand McNally
Made in U.S.A.

# Antarctica Blank Map

Name: ______________________ Date: ______________________

See page **224** in Goode's World Atlas (21e) for a full color map.

Label the following **features** on the blank political map of Antarctica.

| | |
|---|---|
| American Highland | Pensacola Mountains |
| Amery Ice Shelf | Queen Maud Land |
| Amundsen Sea | Queen Maud Mountains |
| Bellingshausen Sea | Ronne Ice Shelf |
| Berkner Island | Ross Ice Shelf |
| Coats Land | Ross Sea |
| Enderby Land | Shackleton Ice Shelf |
| Executive Committee Range | South Pole |
| Filchner Ice Shelf | Southern Ocean |
| Framnes Mountains | Victoria Land |
| Horlick Mountains | Vinson Massif |
| Mt Erebus | Weddell Sea |
| Mt Markham | Whitmore Mountains |
| Mt Sabine | Wilkes Land |
| Mt Sidley | |

Koala, Australia *Corbis Images*

Sydney, Australia *Corbis Digital Stock*

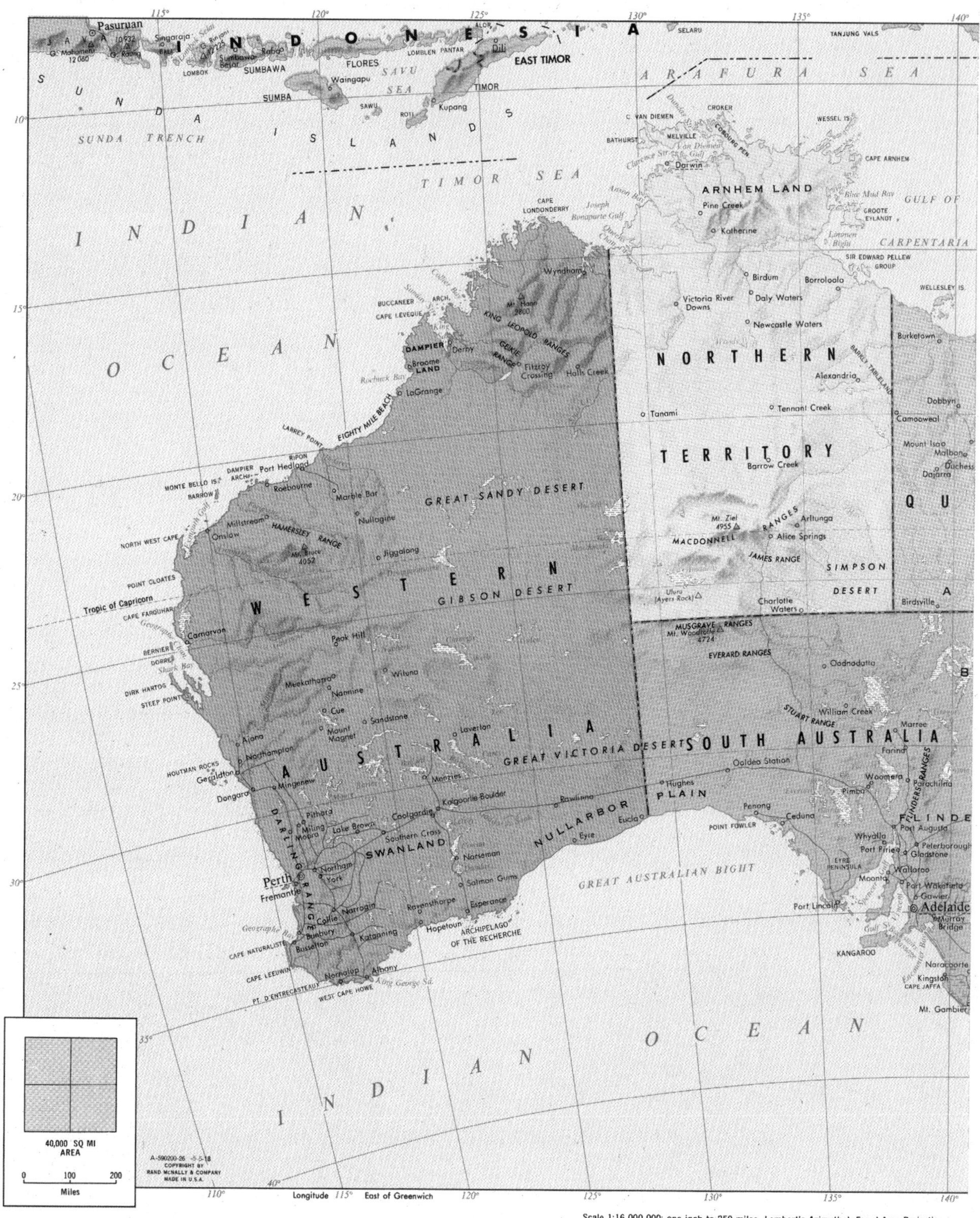


Scale 1:16 000 000; one inch to 250 miles. Lambert's Azimuthal, Equal Area Projection
Elevations and depressions are given in feet

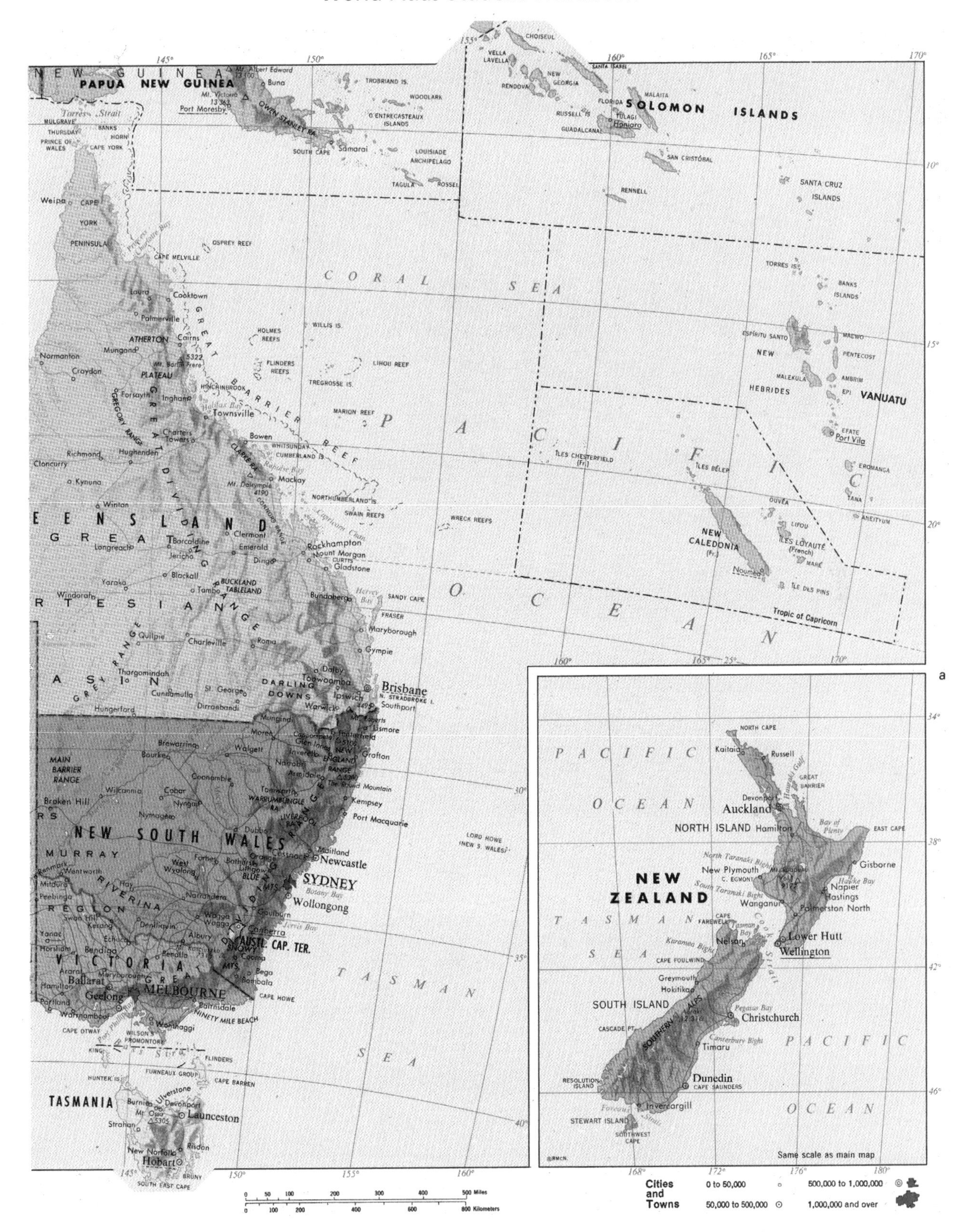
PAPUA NEW GUINEA
Port Moresby
Torres Strait
CAPE YORK
Weipa
CAPE YORK PENINSULA
CAPE MELVILLE
OSPREY REEF
CORAL SEA
Cooktown
Palmerville
ATHERTON
Cairns
Mungana
Normanton
Croydon
PLATEAU
Forsayth
Ingham
Townsville
GREAT BARRIER REEF
WILLIS IS.
HOLMES REEFS
FLINDERS REEFS
LIHOU REEF
TREGROSSE IS.
MARION REEF
Bowen
Charters Towers
Hughenden
Richmond
Cloncurry
Kynuna
Mackay
Winton
QUEENSLAND
GREAT DIVIDING RANGE
Clermont
Emerald
Barcaldine
Longreach
Jericho
Rockhampton
Mount Morgan
Gladstone
Blackall
Tambo
Yaraka
Windorah
Bundaberg
Hervey Bay
SANDY CAPE
FRASER
Maryborough
Gympie
Quilpie
Charleville
Roma
Thargomindah
Dalby
Toowoomba
Brisbane
DARLING DOWNS
Ipswich
Southport
Warwick
Cunnamulla
St. George
Dirranbandi
Hungerford
GREAT ARTESIAN BASIN
WRECK REEFS
SOLOMON ISLANDS
CHOISEUL
VELLA LAVELLA
NEW GEORGIA
RENDOVA
SANTA ISABEL
MALAITA
FLORIDA
RUSSELL IS.
GUADALCANAL
Honiara
SAN CRISTÓBAL
RENNELL
SANTA CRUZ ISLANDS
TORRES IS.
BANKS ISLANDS
ESPÍRITU SANTO
NEW HEBRIDES
MAEWO
PENTECOST
MALEKULA
AMBRIM
EPI
VANUATU
EFATE
Port Vila
EROMANGA
TANA
ANEITYUM
WOODLARK
D'ENTRECASTEAUX ISLANDS
TROBRIAND IS.
LOUISIADE ARCHIPELAGO
TAGULA
ROSSEL
SOUTH CAPE
Samarai
PACIFIC OCEAN
ÎLES CHESTERFIELD (Fr.)
ÎLES BÉLEP
OUVÉA
LIFOU
ÎLES LOYAUTÉ (French)
MARÉ
NEW CALEDONIA (Fr.)
Nouméa
ÎLE DES PINS
Tropic of Capricorn
MAIN BARRIER RANGE
Broken Hill
Wilcannia
Bourke
Brewarrina
Walgett
Moree
Cobar
Nyngan
Coonamble
Narrabri
Inverell
Glen Innes
Armidale
Tamworth
Lismore
Grafton
Kempsey
Port Macquarie
Dubbo
NEW SOUTH WALES
MURRAY
Wentworth
Mildura
Hay
Forbes
Bathurst
Lithgow
BLUE MTS.
Maitland
Newcastle
SYDNEY
Botany Bay
Wollongong
RIVERINA
REGION
Narrandera
Wagga Wagga
Goulburn
Canberra
AUSTL. CAP. TER.
Deniliquin
Albury
Echuca
Bendigo
Swan Hill
Kerang
VICTORIA
Ballarat
Geelong
MELBOURNE
Hamilton
Portland
Warrnambool
CAPE OTWAY
WILSON'S PROMONTORY
Wonthaggi
Bairnsdale
NINETY MILE BEACH
Cooma
Bega
Bombala
CAPE HOWE
LORD HOWE (NEW S. WALES)
TASMAN SEA
KING
FLINDERS
FURNEAUX GROUP
CAPE BARREN
HUNTER IS.
TASMANIA
Burnie
Devonport
Ulverstone
Launceston
Strahan
New Norfolk
Hobart
SOUTH EAST CAPE
NORTH CAPE
Kaitaia
Russell
GREAT BARRIER
Devonport
Auckland
NORTH ISLAND
Hamilton
Bay of Plenty
EAST CAPE
North Taranaki Bight
New Plymouth
C. EGMONT
Gisborne
Hawke Bay
Napier
Hastings
NEW ZEALAND
South Taranaki Bight
Wanganui
Palmerston North
TASMAN SEA
CAPE FAREWELL
Tasman Bay
Nelson
Karamea Bight
Lower Hutt
Wellington
Cook Strait
CAPE FOULWIND
Greymouth
Hokitika
SOUTH ISLAND
SOUTHERN ALPS
Pegasus Bay
Christchurch
CASCADE PT.
Canterbury Bight
Timaru
PACIFIC OCEAN
RESOLUTION ISLAND
Dunedin
CAPE SAUNDERS
Foveaux Strait
Invercargill
STEWART ISLAND
SOUTHWEST CAPE
Same scale as main map
Cities and Towns
0 to 50,000
50,000 to 500,000
500,000 to 1,000,000
1,000,000 and over
Miles
Kilometers

Dorrigo National Park, New South Wales, Australia *Digital Vision*

## *Australia Political Map*

Name: ______________________________ Date: ______________________________

See pages **218-219** in Goode's World Atlas (21e) for a full color map.

List a **city** for each of the following letters.

| | |
|---|---|
| A ______________________ | M ______________________ |
| B ______________________ | N ______________________ |
| C ______________________ | O ______________________ |
| D ______________________ | P ______________________ |
| E ______________________ | Q ______________________ |
| F ______________________ | R ______________________ |
| G ______________________ | S ______________________ |
| H ______________________ | T ______________________ |
| I ______________________ | V ______________________ |
| J ______________________ | W ______________________ |
| K ______________________ | Y ______________________ |
| L ______________________ | |

List all of the states or territories through which the Tropic of Capricorn passes.

List all of the states or territories that are situated south of the Tropic of Capricorn.

What state or territory occupies the largest areal extent? Which is the smallest?

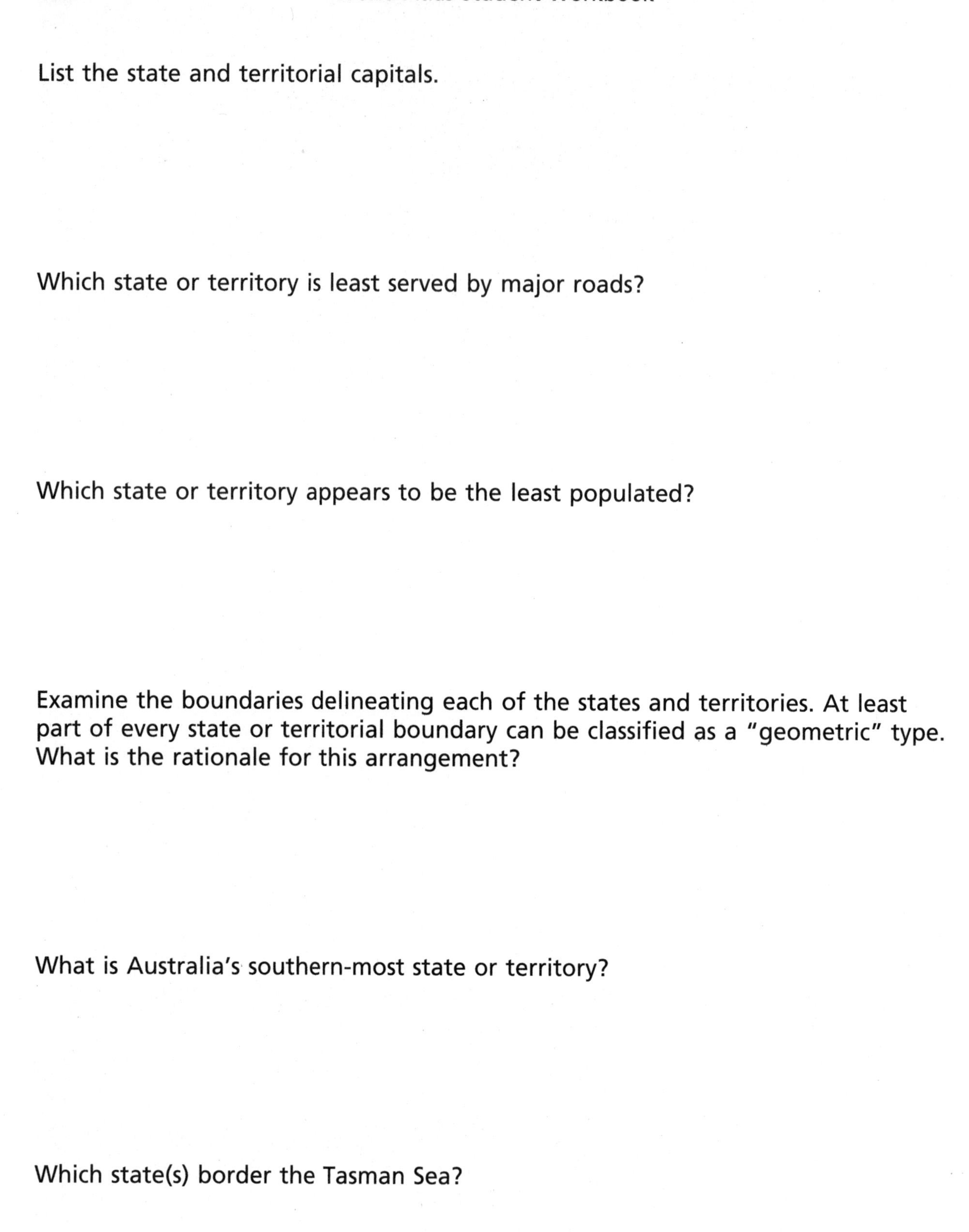

List the state and territorial capitals.

Which state or territory is least served by major roads?

Which state or territory appears to be the least populated?

Examine the boundaries delineating each of the states and territories. At least part of every state or territorial boundary can be classified as a "geometric" type. What is the rationale for this arrangement?

What is Australia's southern-most state or territory?

Which state(s) border the Tasman Sea?

Great Barrier Reef, Australia *Corbis Digital Stock*

Scale 1:16 000 000; one inch to 250 miles. Lambert's Azimuthal, Equal Area Projection
Elevations and depressions are given in feet

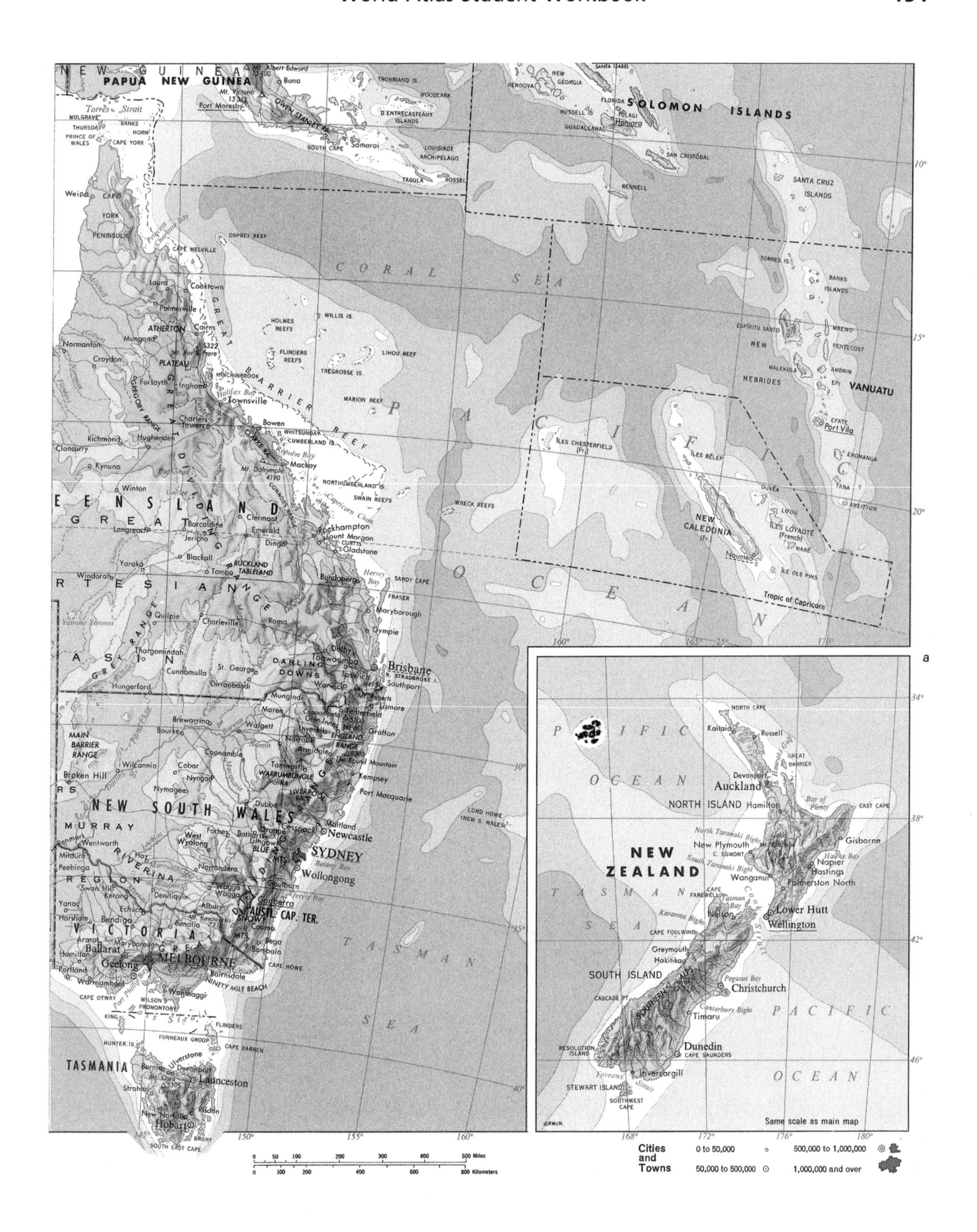
PAPUA NEW GUINEA
Port Moresby
Torres Strait
CAPE YORK
Weipa
CAPE YORK PENINSULA
CAPE MELVILLE
SOLOMON ISLANDS
Honiara
GUADALCANAL
SAN CRISTÓBAL
RENNELL
SANTA CRUZ ISLANDS
LOUISIADE ARCHIPELAGO
D'ENTRECASTEAUX ISLANDS
TROBRIAND IS.
CORAL SEA
Cooktown
Cairns
ATHERTON PLATEAU
Normanton
Croydon
Townsville
GREAT BARRIER REEF
HOLMES REEFS
WILLIS IS.
FLINDERS REEFS
LIHOU REEF
TREGROSSE IS.
MARION REEF
Bowen
WHITSUNDAY I.
CUMBERLAND IS.
Mackay
Charters Towers
Hughenden
Richmond
Cloncurry
Kynuna
Winton
NORTHUMBERLAND IS.
SWAIN REEFS
WRECK REEFS
Clermont
Rockhampton
Mount Morgan
Gladstone
Emerald
Barcaldine
Longreach
Jericho
Blackall
Tambo
BUCKLAND TABLELAND
Bundaberg
SANDY CAPE
FRASER
Maryborough
Gympie
Charleville
Roma
Quilpie
Windorah
Thargomindah
Cunnamulla
St. George
Dirranbandi
Hungerford
DARLING DOWNS
Toowoomba
Brisbane
N. STRADBROKE I.
Southport
Lismore
Grafton
Moree
Walgett
Bourke
Brewarrina
MAIN BARRIER RANGE
Broken Hill
Wilcannia
Cobar
Nyngan
Nymagee
Coonamble
NEW ENGLAND RANGE
Armidale
Tamworth
WARRUMBUNGLE RA.
Kempsey
Port Macquarie
LIVERPOOL RA.
Dubbo
NEW SOUTH WALES
LORD HOWE (NEW S. WALES)
Maitland
Cessnock
Newcastle
SYDNEY
Botany Bay
Wollongong
BLUE MTS.
Bathurst
Orange
Forbes
West Wyalong
MURRAY
RIVERINA
Wentworth
Mildura
Hay
Narrandera
Wagga Wagga
Goulburn
Deniliquin
Albury
Canberra
AUSTL. CAP. TER.
SNOWY MTS.
Cooma
Bega
Bombala
VICTORIA
Bendigo
Echuca
Benalla
Horsham
Ararat
Ballarat
Hamilton
Geelong
MELBOURNE
Portland
Warrnambool
Port Phillip Bay
CAPE OTWAY
Bairnsdale
NINETY MILE BEACH
CAPE HOWE
Wonthaggi
WILSON'S PROMONTORY
KING I.
Bass Strait
FLINDERS I.
FURNEAUX GROUP
CAPE BARREN
HUNTER IS.
TASMANIA
Burnie
Ulverstone
Devonport
Launceston
Strahan
New Norfolk
Hobart
BRUNY I.
SOUTH EAST CAPE
TASMAN SEA
PACIFIC OCEAN
ÎLES CHESTERFIELD (Fr.)
ÎLES BÉLEP
NEW CALEDONIA (Fr.)
Nouméa
ÎLE DES PINS
OUVÉA
LIFOU
ÎLES LOYAUTÉ (French)
MARÉ
Tropic of Capricorn
TORRES IS.
BANKS ISLANDS
ESPÍRITU SANTO
NEW HEBRIDES
MAEWO
PENTECOST
MALEKULA
AMBRIM
EPI
VANUATU
EFATE
Port Vila
EROMANGA
TANA
ANEITYUM
NORTH CAPE
Kaitaia
Russell
GREAT BARRIER I.
Devonport
Auckland
NORTH ISLAND
Hamilton
Bay of Plenty
EAST CAPE
Gisborne
North Taranaki Bight
New Plymouth
C. EGMONT
South Taranaki Bight
Hawke Bay
Napier
Hastings
Wanganui
Palmerston North
NEW ZEALAND
CAPE FAREWELL
Tasman Bay
Nelson
Lower Hutt
Wellington
Karamea Bight
CAPE FOULWIND
Greymouth
Hokitika
SOUTH ISLAND
SOUTHERN ALPS
Pegasus Bay
Christchurch
CASCADE PT.
Canterbury Bight
Timaru
RESOLUTION ISLAND
Dunedin
CAPE SAUNDERS
Foveaux Strait
Invercargill
STEWART ISLAND
SOUTHWEST CAPE
Same scale as main map
0 50 100 200 300 400 500 Miles
0 100 200 400 600 800 Kilometers
Cities and Towns
0 to 50,000
50,000 to 500,000
500,000 to 1,000,000
1,000,000 and over

Perth, Australia *Corbis Digital Stock*

## *Australia Physical Map*

Name: ______________________ Date: ______________________

See pages **220-221** in Goode's World Atlas (21e) for a full color map.

What are the principal **mountain ranges** on the continent?

______________________

______________________

______________________

What are the major **rivers** on the continent?

______________________

______________________

______________________

List the state(s) or territory(s) where you find each of the following **features**.

Great Victorian Desert ______________________
Great Dividing Range ______________________
Gulf of Carpentaria ______________________
Great Sandy Desert ______________________
Macdonnell Range ______________________
Mount Bruce ______________________
Gibson Desert ______________________
Eighty Mile Beach ______________________
Hamersley Range ______________________
Nullarbor Plain ______________________
Mount Kosciuszko ______________________
Bass Strait ______________________
Great Barrier Reef ______________________
Darling Range ______________________
Mount Woodroffe ______________________
Ninety Mile Beach ______________________
Atherton Plateau ______________________
Gray Range ______________________
Simpson Desert ______________________
Great Artesian Basin ______________________
Snowy Mountains ______________________
Mount Ziel ______________________

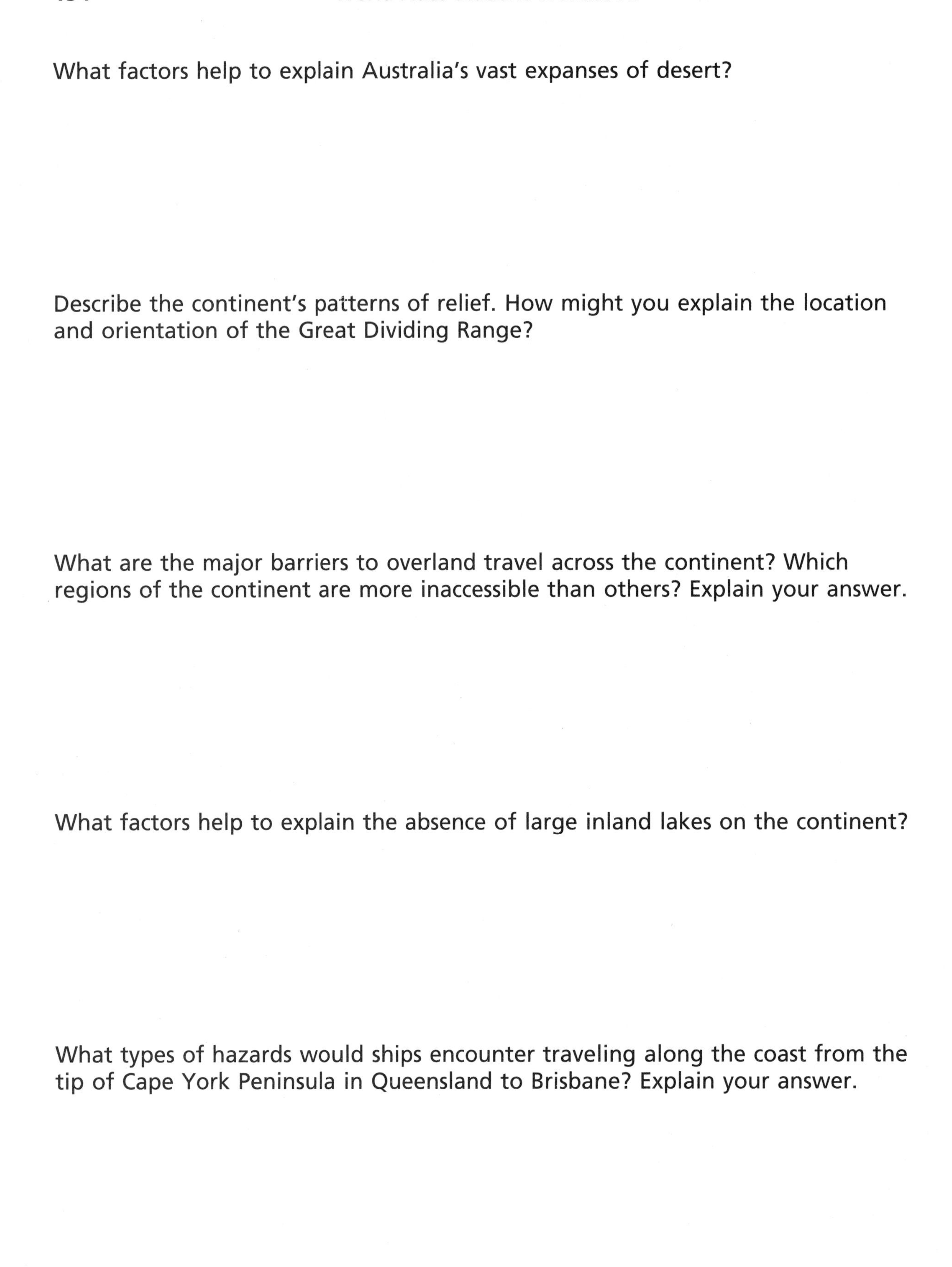

What factors help to explain Australia's vast expanses of desert?

Describe the continent's patterns of relief. How might you explain the location and orientation of the Great Dividing Range?

What are the major barriers to overland travel across the continent? Which regions of the continent are more inaccessible than others? Explain your answer.

What factors help to explain the absence of large inland lakes on the continent?

What types of hazards would ships encounter traveling along the coast from the tip of Cape York Peninsula in Queensland to Brisbane? Explain your answer.

Ayers Rock (Uluru), Australia *Corbis Digital Stock*

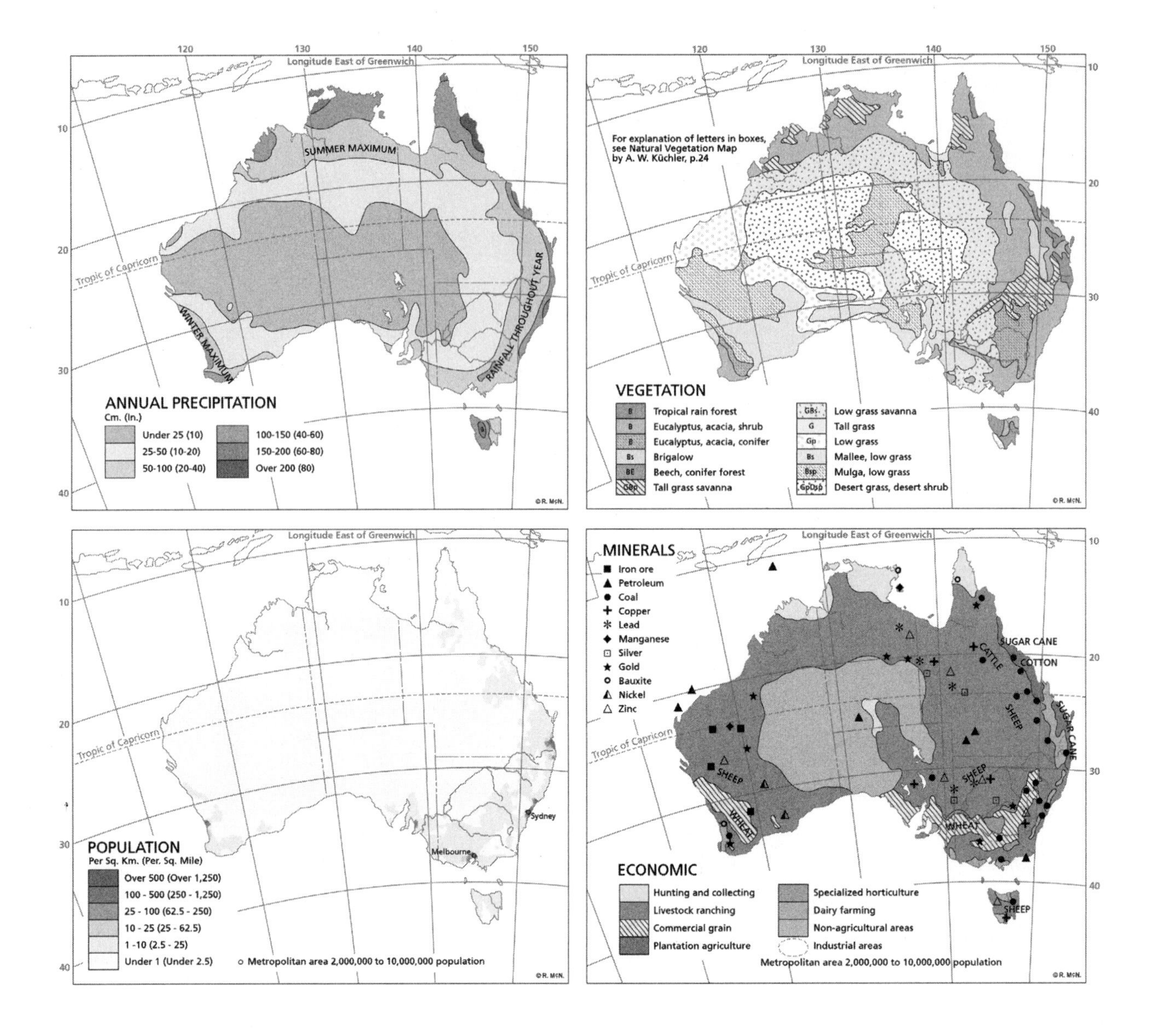
ANNUAL PRECIPITATION
Cm. (In.)
Under 25 (10)
25-50 (10-20)
50-100 (20-40)
100-150 (40-60)
150-200 (60-80)
Over 200 (80)
SUMMER MAXIMUM
WINTER MAXIMUM
RAINFALL THROUGHOUT YEAR
Longitude East of Greenwich
Tropic of Capricorn
VEGETATION
For explanation of letters in boxes, see Natural Vegetation Map by A. W. Küchler, p.24
Tropical rain forest
Eucalyptus, acacia, shrub
Eucalyptus, acacia, conifer
Brigalow
Beech, conifer forest
Tall grass savanna
Low grass savanna
Tall grass
Low grass
Mallee, low grass
Mulga, low grass
Desert grass, desert shrub
POPULATION
Per Sq. Km. (Per. Sq. Mile)
Over 500 (Over 1,250)
100 - 500 (250 - 1,250)
25 - 100 (62.5 - 250)
10 - 25 (25 - 62.5)
1 -10 (2.5 - 25)
Under 1 (Under 2.5)
Metropolitan area 2,000,000 to 10,000,000 population
Sydney
Melbourne
MINERALS
Iron ore
Petroleum
Coal
Copper
Lead
Manganese
Silver
Gold
Bauxite
Nickel
Zinc
SUGAR CANE
COTTON
CATTLE
SHEEP
WHEAT
ECONOMIC
Hunting and collecting
Livestock ranching
Commercial grain
Plantation agriculture
Specialized horticulture
Dairy farming
Non-agricultural areas
Industrial areas
Metropolitan area 2,000,000 to 10,000,000 population
©R. McN.

## *Australia Thematic Maps*

Name: ______________________ Date: ______________________

See page **216** in Goode's World Atlas (21e) for full color maps.

Examine Australia's population distribution. What pattern emerges? What factors may have contributed to this spatial pattern?

Given Australia's limited surface water resources, what activity accounts for the greatest percentage of water use on the continent? What activity is the second largest consumer of water?

Where does plantation agriculture occur on the continent? Why is it prevalent in this particular region?

What regions experience the highest amounts of annual precipitation on the continent? Where do the least amounts occur? What factors contribute to this pattern?

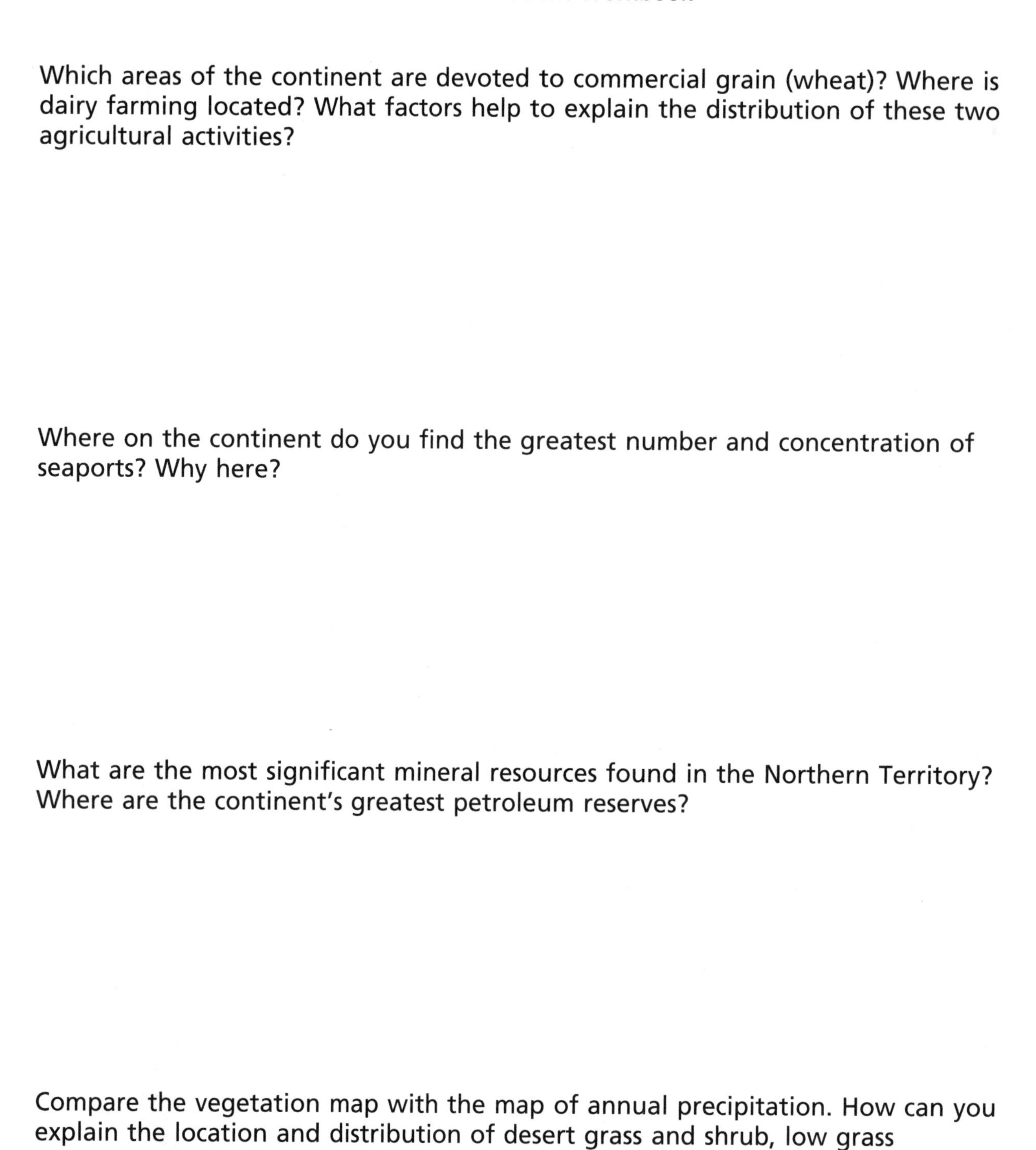

Which areas of the continent are devoted to commercial grain (wheat)? Where is dairy farming located? What factors help to explain the distribution of these two agricultural activities?

Where on the continent do you find the greatest number and concentration of seaports? Why here?

What are the most significant mineral resources found in the Northern Territory? Where are the continent's greatest petroleum reserves?

Compare the vegetation map with the map of annual precipitation. How can you explain the location and distribution of desert grass and shrub, low grass savanna, and eucalyptus – acacia – scrub?

Australia *Corbis Digital Stock*

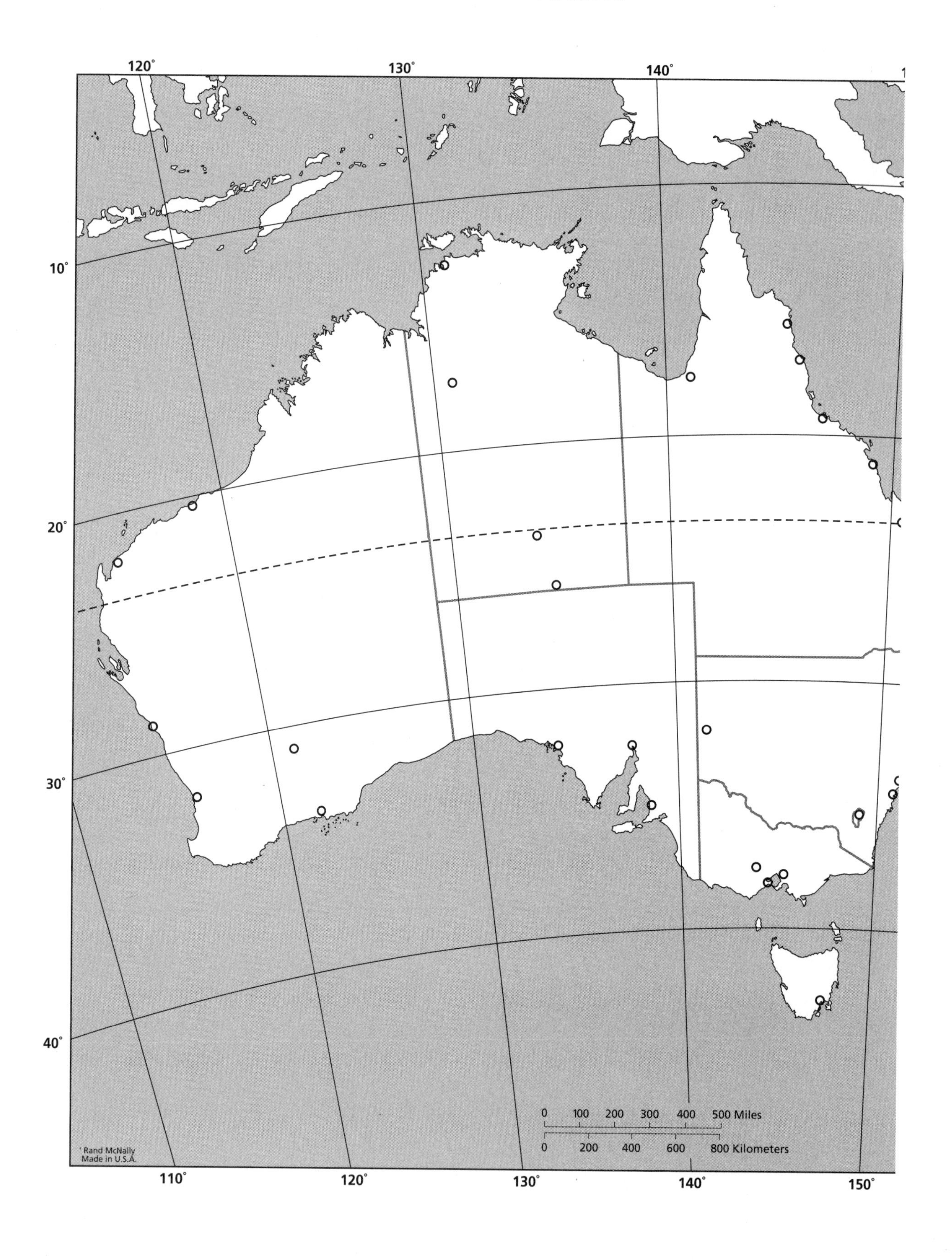
120°
130°
140°
10°
20°
30°
40°
0 100 200 300 400 500 Miles
0 200 400 600 800 Kilometers
© Rand McNally
Made in U.S.A.
110°
120°
130°
140°
150°

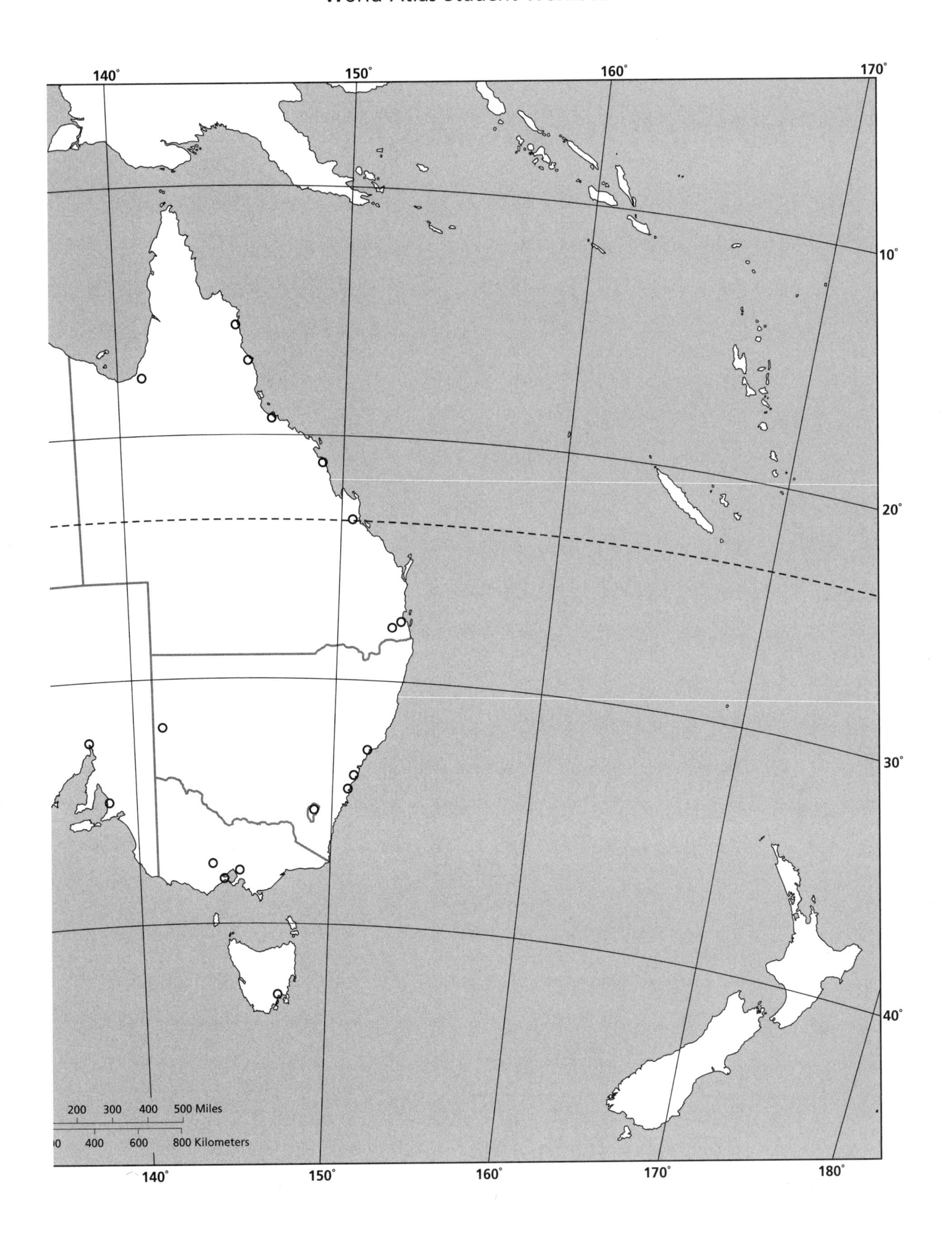

140°
150°
160°
170°
10°
20°
30°
40°
200 300 400 500 Miles
0 400 600 800 Kilometers
140°
150°
160°
170°
180°

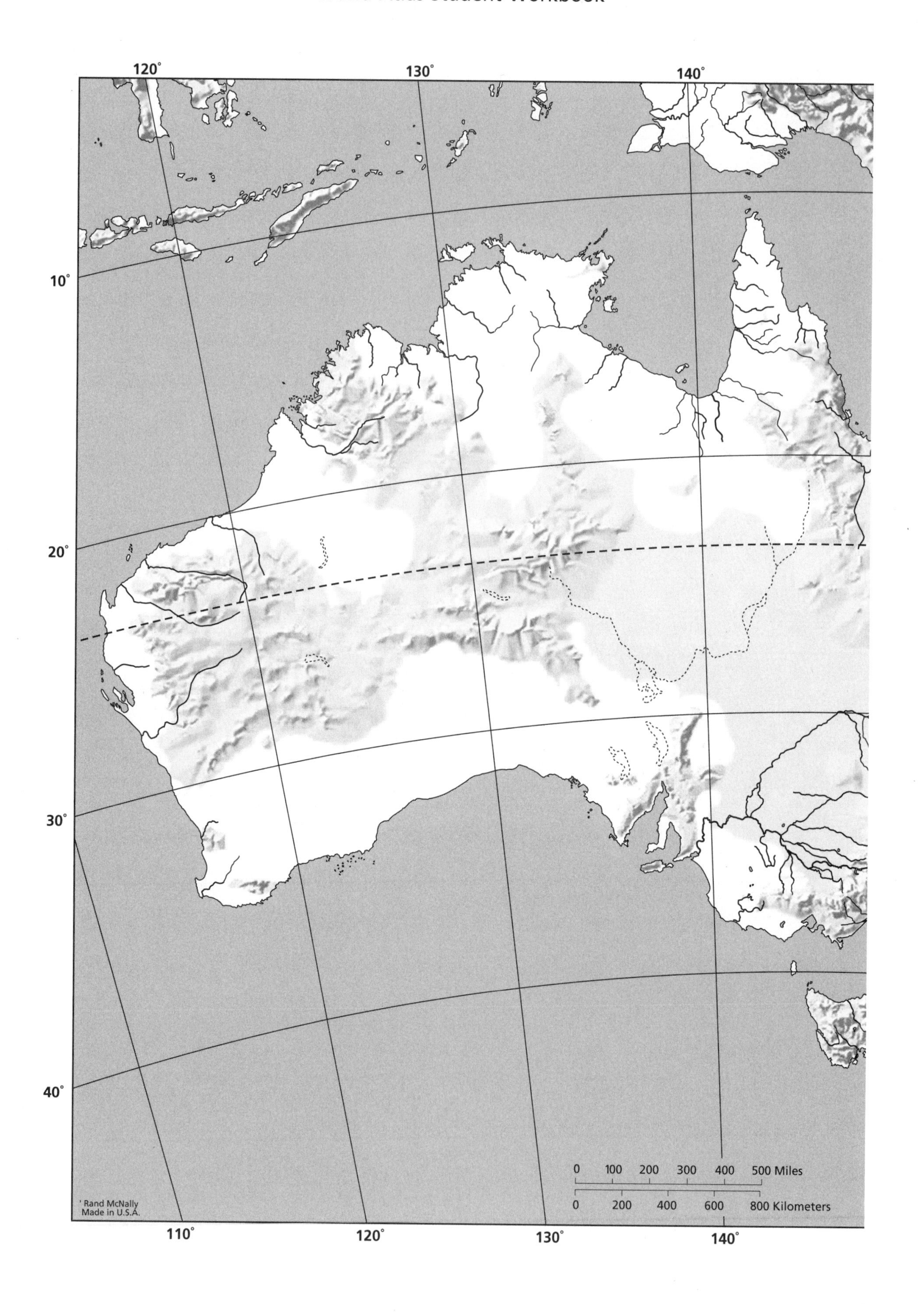
120°
130°
140°
10°
20°
30°
40°
110°
120°
130°
140°
0 100 200 300 400 500 Miles
0 200 400 600 800 Kilometers
Rand McNally
Made in U.S.A.

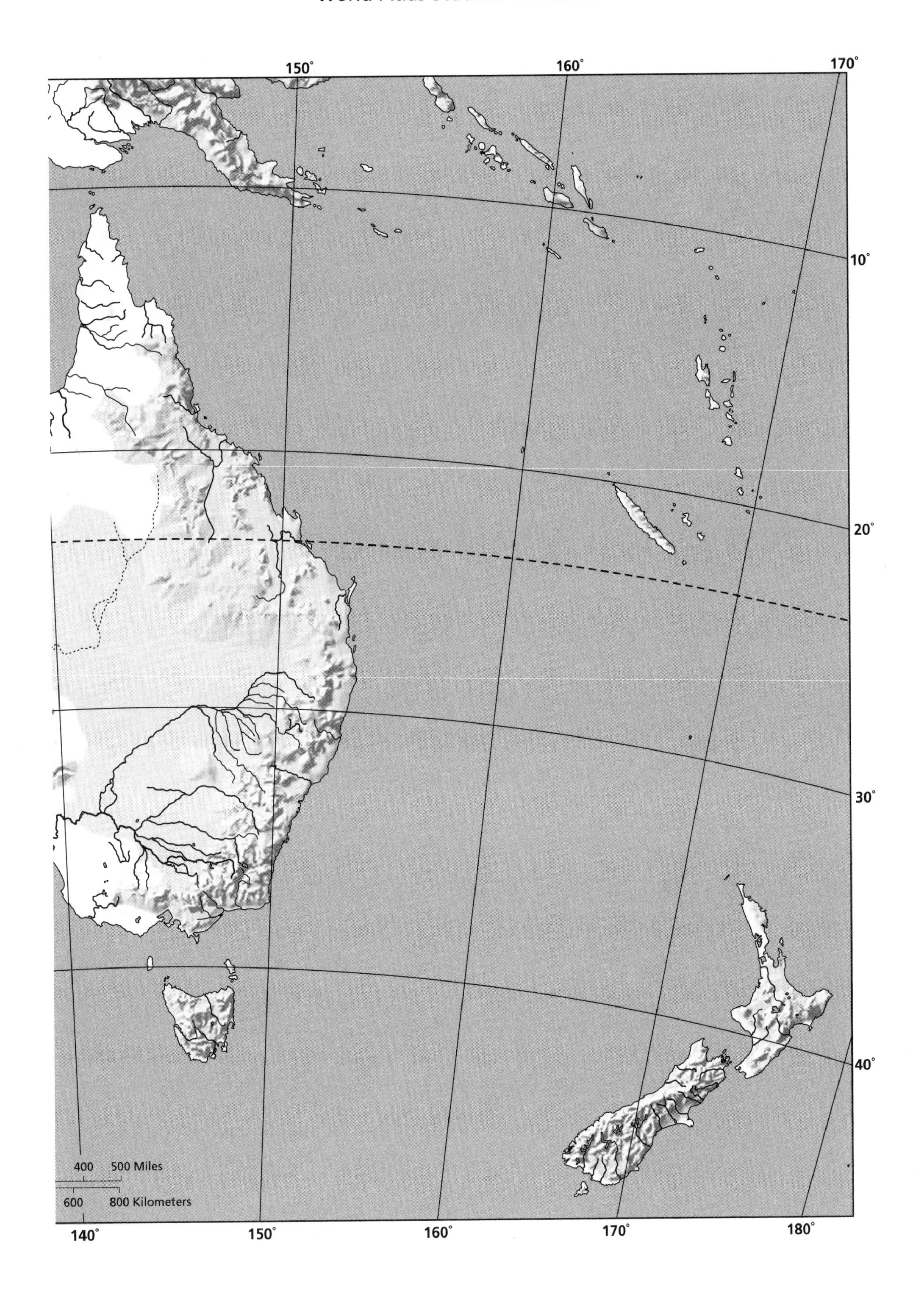
150°
160°
170°
10°
20°
30°
40°
400 500 Miles
600 800 Kilometers
140°
150°
160°
170°
180°

Perth, Australia *Corbis Digital Stock*

## *Australia Blank Maps (Political / Physical)*

Name: ______________________ Date: ______________________

See page(s) **218-221** in Goode's World Atlas (21e) for full color maps.

### Political Map

Label the following **states and territories** on the blank political map of Australia.

| | |
|---|---|
| Australian Capital Territory<br>New South Wales<br>Northern Territory<br>Queensland | South Australia<br>Tasmania<br>Victoria<br>Western Australia |

Label the following **cities** on the blank political map of Australia.

| | |
|---|---|
| Adelaide<br>Alice Springs<br>Ballarat<br>Brisbane<br>Broken Hill<br>Cairns<br>Canberra<br>Ceduna<br>Charlotte Waters<br>Cooktown<br>Coolgardie<br>Darwin<br>Esperance<br>Geelong<br>Geraldton | Holbart<br>Ipswich<br>Mackay<br>Melbourne<br>Newcastle<br>Normanton<br>Onslow<br>Perth<br>Port Augusta<br>Port Hedland<br>Rockhampton<br>Sydney<br>Townsville<br>Victoria River Downs<br>Wollongong |

## Physical Map

Label the following **features** on the blank physical map of Australia.

### *Mountain Ranges*

| | |
|---|---|
| Darling Range<br>Everard Ranges<br>Flinders Ranges<br>Geikie Range<br>Great Dividing range<br>Gregory Range<br>Grey Range | Hamersley Range<br>James Range<br>King Leopold Ranges<br>Macdonnell Range<br>Main Barrier Range<br>Musgrave Ranges<br>Stuart Range |

### *Rivers*

| | |
|---|---|
| Ashburton<br>Barwon<br>Belyando<br>Billabong<br>Daly<br>Darling<br>Dawson<br>DeGrey<br>Finke<br>Fitzroy<br>Fortescue | Gilbert<br>Macquarie<br>Mitchell<br>Murchison<br>Murray<br>Murrumbidgee<br>Namoi<br>Roper<br>Thomson<br>Victoria |

### *Bodies of Water*

| | |
|---|---|
| Arafura Sea<br>Botany Bay<br>Charlotte Bay<br>Coral Sea<br>Exmouth Gulf<br>Great Australian Bight<br>Gulf of Carpentaria<br>Halifax Bay<br>Hervey Bay | Indian Ocean<br>Jervis Bay<br>Joseph Bonaparte Gulf<br>Pacific Ocean<br>Repulse Bay<br>Shark Bay<br>Tasman Sea<br>Timor Sea<br>Van Diemen Gulf |

### *Deserts*

| | |
|---|---|
| Gibson<br>Great Sandy | Great Victoria<br>Simpson |

Mt Kilimanjaro, Tanzania *Corbis Digital Stock*

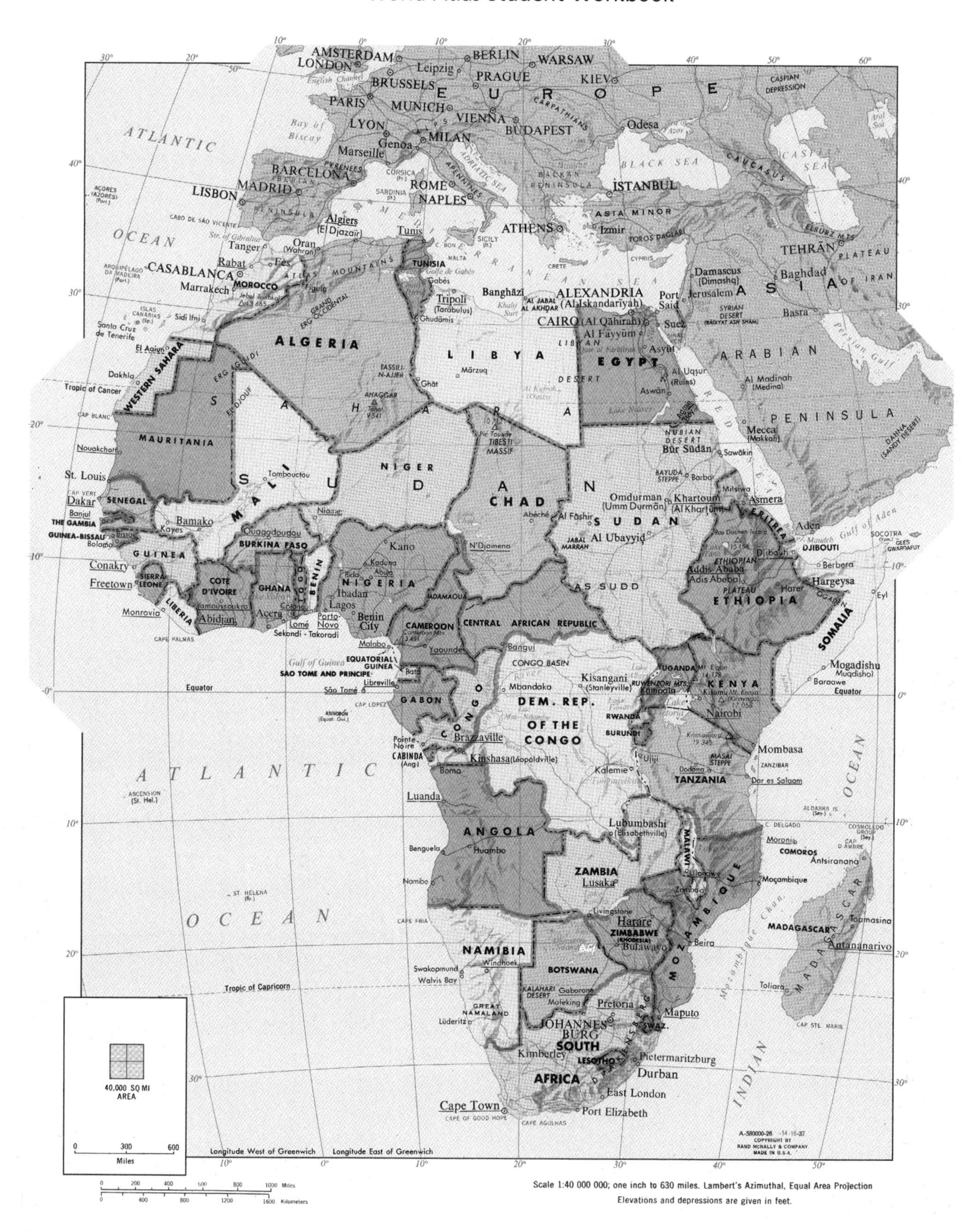

AMSTERDAM
LONDON
BERLIN
WARSAW
Leipzig
PRAGUE
BRUSSELS
KIEV
E U R O P E
PARIS
MUNICH
VIENNA
BUDAPEST
Odesa
LYON
Genoa
MILAN
Marseille
Bay of Biscay
ATLANTIC
OCEAN
BARCELONA
MADRID
LISBON
ROME
NAPLES
CORSICA
SARDINIA
BLACK SEA
ISTANBUL
CASPIAN DEPRESSION
CASPIAN SEA
Aral Sea
ASIA MINOR
Izmir
ATHENS
Algiers
(El Djazaïr)
Tunis
SICILY
Oran
(Wahran)
Tanger
Rabat
Fès
CASABLANCA
MOROCCO
Marrakech
ATLAS MOUNTAINS
TUNISIA
Gabès
MALTA
CRETE
CYPRUS
Damascus
(Dimashq)
Jerusalem
Baghdad
TEHRĀN
PLATEAU OF IRAN
A S I A
Basra
Tripoli
(Tarabulus)
Ghudāmis
Banghāzī
ALEXANDRIA
(Al Iskandarīyah)
Port Said
Suez
CAIRO (Al Qāhirah)
Al Fayyūm
Asyūt
SYRIAN DESERT
ARABIAN PENINSULA
Sidi Ifni
Santa Cruz de Tenerife
El Aaiún
ALGERIA
LIBYA
EGYPT
Mārzuq
Ghāt
Al Uqsur
Aswān
Dakhla
Tropic of Cancer
WESTERN SAHARA
S A H A R A
LIBYAN DESERT
Al Madīnah
(Medina)
Mecca
(Makkah)
NUBIAN DESERT
Būr Sūdān
Sawākin
MAURITANIA
Nouakchott
St. Louis
Dakar
SENEGAL
Banjul
THE GAMBIA
GUINEA-BISSAU
Bissau
Bolama
MALI
Tombouctou
NIGER
CHAD
SUDAN
S U D A N
Omdurman
(Umm Durmān)
Khartoum
(Al Khartūm)
Asmera
ERITREA
Aden
Gulf of Aden
SOCOTRA
Bamako
Kayes
Niamey
Ouagadougou
BURKINA FASO
Kano
N'Djamena
Abéché
Al Fāshir
Al Ubayyid
DJIBOUTI
Djibouti
GUINEA
Conakry
Freetown
SIERRA LEONE
COTE D'IVOIRE
GHANA
TOGO
BENIN
NIGERIA
Kaduna
Abuja
Ibadan
Lagos
Benin City
ETHIOPIA
Addis Ababa
ETHIOPIAN PLATEAU
Harer
Berbera
Hargeysa
Monrovia
LIBERIA
Abidjan
Yamoussoukro
Accra
Lomé
Porto-Novo
Sekondi - Takoradi
CAMEROON
CENTRAL AFRICAN REPUBLIC
AS SUDD
SOMALIA
Eyl
Malabo
Yaoundé
Bangui
Gulf of Guinea
EQUATORIAL GUINEA
SAO TOME AND PRINCIPE
Libreville
São Tomé
Equator
CONGO BASIN
Mbandaka
Kisangani
(Stanleyville)
UGANDA
Kampala
KENYA
Nairobi
Mogadishu
(Muqdisho)
GABON
CONGO
DEM. REP. OF THE CONGO
RWANDA
BURUNDI
Pointe Noire
Brazzaville
CABINDA
Kinshasa (Léopoldville)
Boma
Mombasa
Kalemie
TANZANIA
Dodoma
Dar es Salaam
ATLANTIC
OCEAN
ASCENSION
(St. Hel.)
Luanda
ANGOLA
Huambo
Benguela
Lubumbashi
(Elisabethville)
MALAWI
COMOROS
Moroni
Antsiranana
ZAMBIA
Lusaka
Namibe
Moçambique
ST. HELENA
Livingstone
Harare
ZIMBABWE
(RHODESIA)
Bulawayo
MOZAMBIQUE
Beira
MADAGASCAR
Toamasina
Antananarivo
NAMIBIA
Windhoek
Swakopmund
Walvis Bay
BOTSWANA
KALAHARI DESERT
Gaborone
Tropic of Capricorn
Toliara
Pretoria
Maputo
GREAT NAMALAND
Lüderitz
JOHANNESBURG
SWAZ.
SOUTH AFRICA
Kimberley
LESOTHO
Pietermaritzburg
Durban
East London
Cape Town
Port Elizabeth
CAPE OF GOOD HOPE
CAPE AGULHAS
INDIAN OCEAN
40,000 SQ MI AREA
Miles
Longitude West of Greenwich
Longitude East of Greenwich
Scale 1:40 000 000; one inch to 630 miles. Lambert's Azimuthal, Equal Area Projection
Elevations and depressions are given in feet.

# *Africa Political Map*

Name: ______________________________ Date: ______________________________

See page **228** in Goode's World Atlas (21e) for a full color map.

List a **country** or major **city** for each of the following letters.

| | |
|---|---|
| A ______________________ | N ______________________ |
| B ______________________ | O ______________________ |
| C ______________________ | P ______________________ |
| D ______________________ | Q ______________________ |
| E ______________________ | R ______________________ |
| F ______________________ | S ______________________ |
| G ______________________ | T ______________________ |
| H ______________________ | U ______________________ |
| I ______________________ | V ______________________ |
| J ______________________ | W ______________________ |
| K ______________________ | Y ______________________ |
| L ______________________ | Z ______________________ |
| M ______________________ | |

List all the countries that border the Equator.

List all the countries along Africa's Atlantic coast.

Which countries in Africa are landlocked?

Which countries border South Africa?

Which countries in Africa border the Mediterranean?

Which countries border the Red Sea?

Which countries are along the Nile River, including the White Nile and Blue Nile?

Which countries border Lake Victoria?

List all the countries on the Arabian Peninsula.

Which countries border Sudan?

Africa *Corbis Digital Stock*

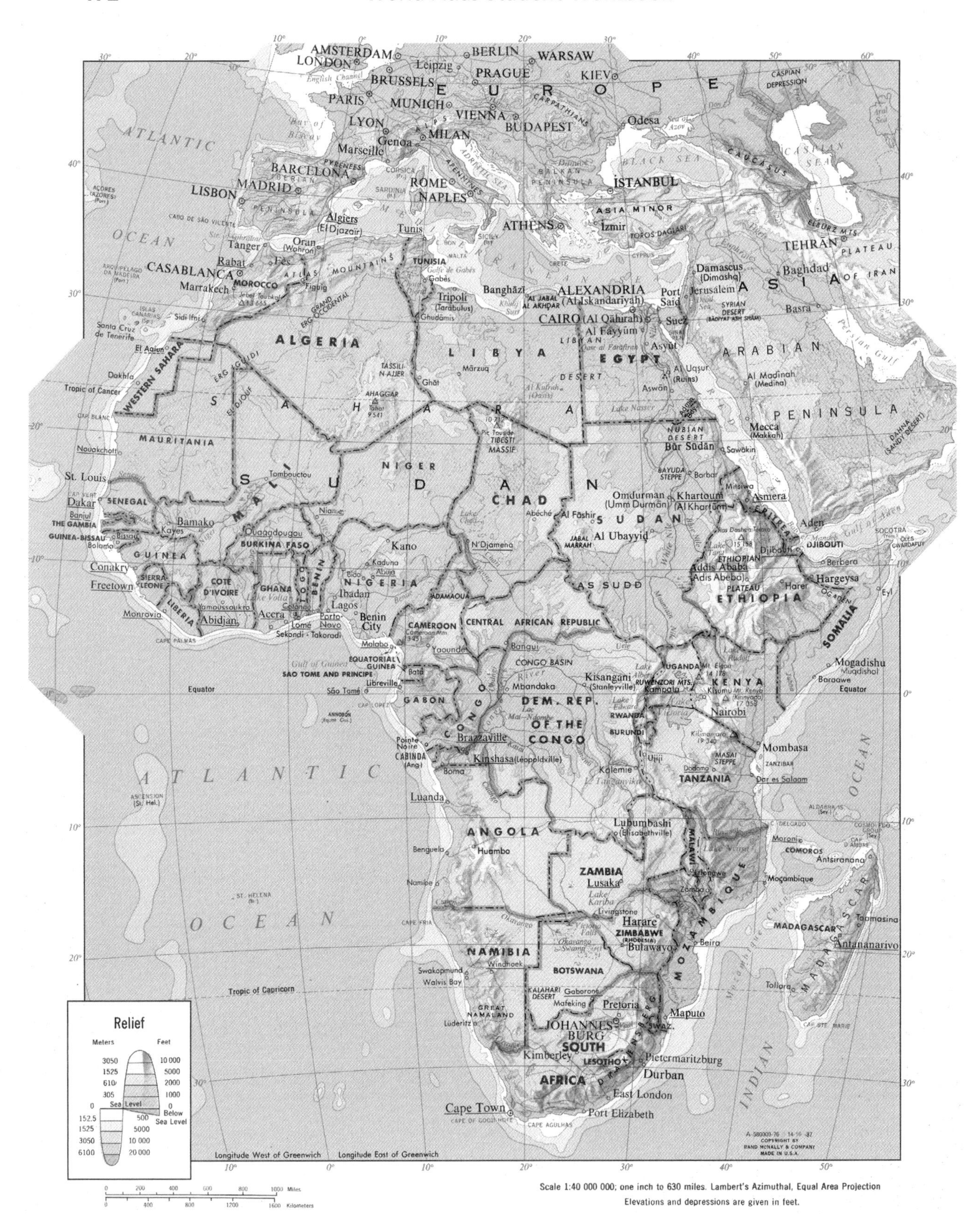


Scale 1:40 000 000; one inch to 630 miles. Lambert's Azimuthal, Equal Area Projection

Elevations and depressions are given in feet.

## *Africa Physical Map*

Name: ______________________ Date: ______________________

See page **227** in Goode's World Atlas (21e) for a full color map.

What are the major **mountain ranges** on the continent?

______________________________

______________________________

______________________________

What are the major **rivers** on the continent?

______________________________

______________________________

______________________________

List the country or countries where you find each of the following **features**.

Lake Chad ______________________
Lake Victoria ______________________
Lake Nyasa ______________________
Lake Tanganyika ______________________
Mount Kenya ______________________
Mount Kilimanjaro ______________________
Congo River ______________________
Niger River ______________________
Nile River ______________________
Sahara Desert ______________________
Atlas Mountains ______________________
Okavango Swamp ______________________
Kalahari Desert ______________________
Lake Albert ______________________
Lake Edward ______________________
Lake Nassar ______________________
Victoria Falls ______________________
Island of Zanzibar ______________________
Cape of Good Hope ______________________
Drakensberg Range ______________________
Vaal River ______________________
Namib Desert ______________________

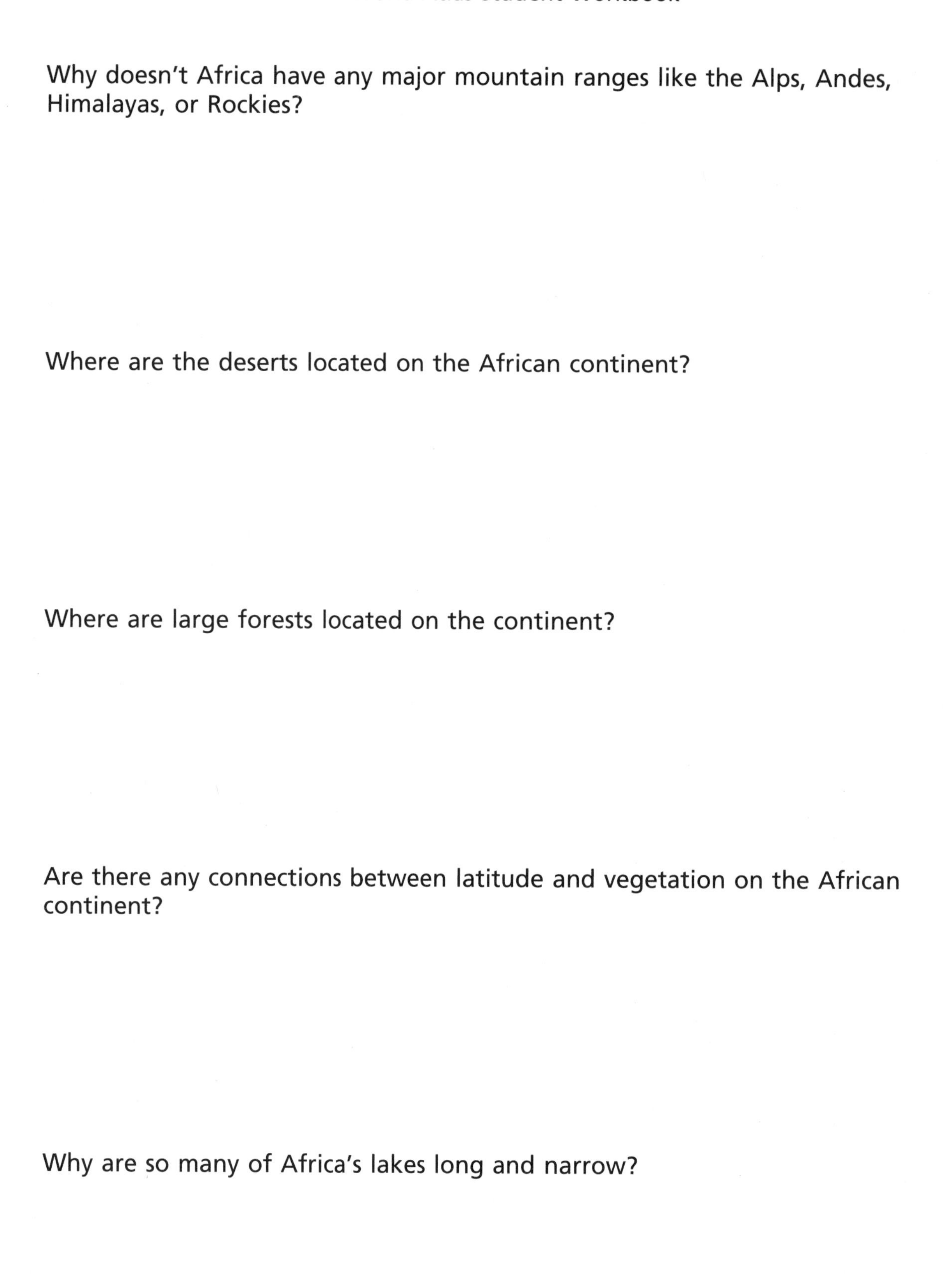

Why doesn't Africa have any major mountain ranges like the Alps, Andes, Himalayas, or Rockies?

Where are the deserts located on the African continent?

Where are large forests located on the continent?

Are there any connections between latitude and vegetation on the African continent?

Why are so many of Africa's lakes long and narrow?

Giza, Egypt *Corbis Digital Stock*

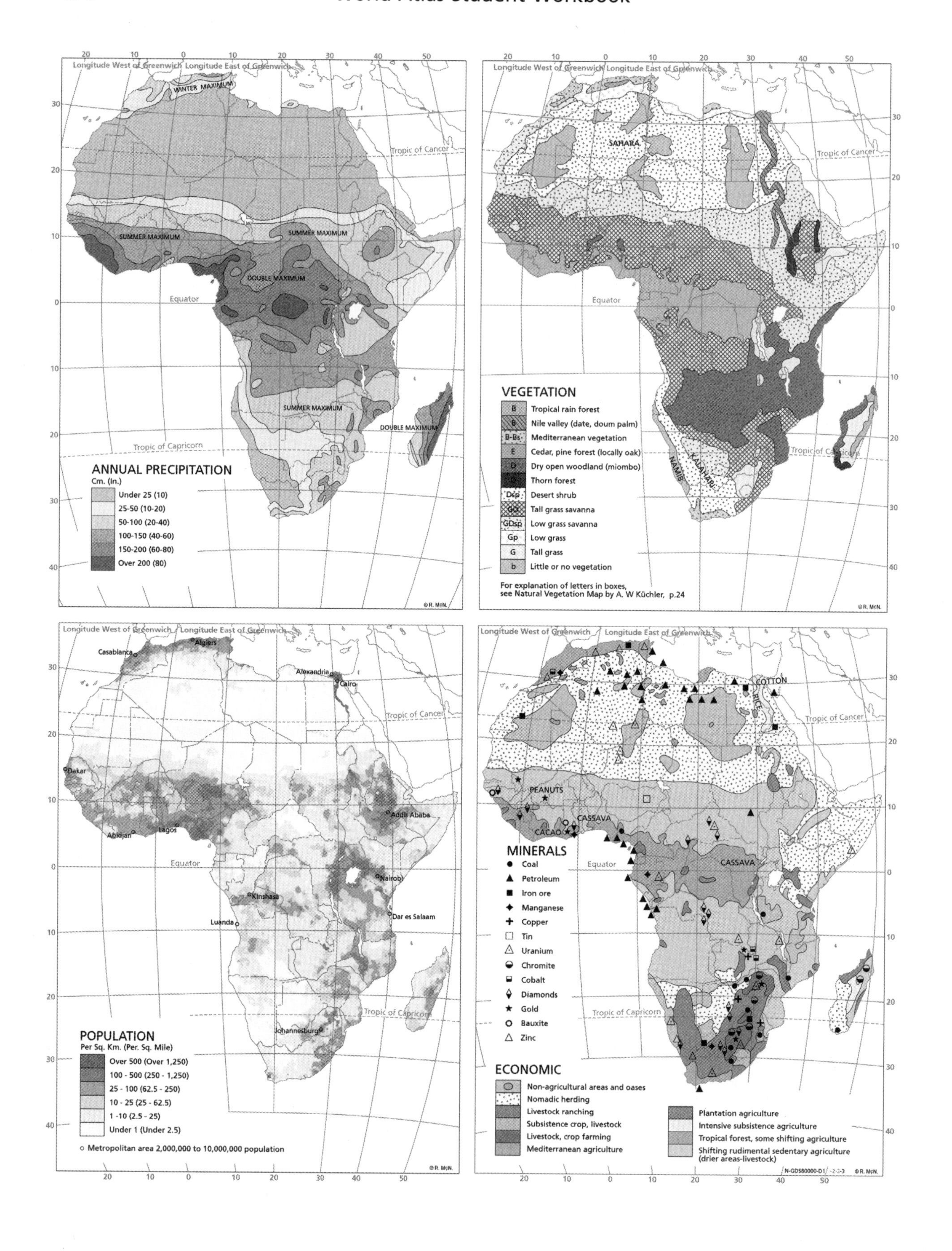

Longitude West of Greenwich Longitude East of Greenwich
WINTER MAXIMUM
SUMMER MAXIMUM
DOUBLE MAXIMUM
Tropic of Cancer
Equator
Tropic of Capricorn
ANNUAL PRECIPITATION
Cm. (In.)
Under 25 (10)
25-50 (10-20)
50-100 (20-40)
100-150 (40-60)
150-200 (60-80)
Over 200 (80)
SAHARA
NAMIB
KALAHARI
VEGETATION
B Tropical rain forest
B Nile valley (date, doum palm)
B-Bs Mediterranean vegetation
E Cedar, pine forest (locally oak)
D Dry open woodland (miombo)
D Thorn forest
Dsp Desert shrub
GD Tall grass savanna
GDsp Low grass savanna
Gp Low grass
G Tall grass
b Little or no vegetation
For explanation of letters in boxes, see Natural Vegetation Map by A. W Küchler, p.24
Casablanca
Algiers
Alexandria
Cairo
Dakar
Abidjan
Lagos
Addis Ababa
Nairobi
Kinshasa
Luanda
Dar es Salaam
Johannesburg
POPULATION
Per Sq. Km. (Per. Sq. Mile)
Over 500 (Over 1,250)
100 - 500 (250 - 1,250)
25 - 100 (62.5 - 250)
10 - 25 (25 - 62.5)
1 -10 (2.5 - 25)
Under 1 (Under 2.5)
Metropolitan area 2,000,000 to 10,000,000 population
COTTON
RICE
PEANUTS
CASSAVA
CACAO
MINERALS
Coal
Petroleum
Iron ore
Manganese
Copper
Tin
Uranium
Chromite
Cobalt
Diamonds
Gold
Bauxite
Zinc
ECONOMIC
Non-agricultural areas and oases
Nomadic herding
Livestock ranching
Subsistence crop, livestock
Livestock, crop farming
Mediterranean agriculture
Plantation agriculture
Intensive subsistence agriculture
Tropical forest, some shifting agriculture
Shifting rudimental sedentary agriculture (drier areas-livestock)
©R. McN.

## *Africa Thematic Maps*

Name: ______________________________ Date: ______________________________

See page **226** in Goode's World Atlas (21e) for full color maps.

Where are the largest population clusters in Africa? What physical geography factors might explain this pattern?

What areas of Africa receive the most annual precipitation?

What similarities are there between Africa's precipitation and vegetation patterns?

What vegetation zones in Africa do you find both north and south of the Equator?

What vegetation areas do you find just outside the boundaries of major deserts?

Where are most of Africa's important minerals located?

Where are the major petroleum-producing regions of Africa?

What vegetation types do you find along the Equator in Africa? Why doesn't the tropical rain forest zone extend across the whole continent?

Why does Ethiopia have different precipitation and vegetation zones than the surrounding countries?

Summarize Africa's agricultural patterns.

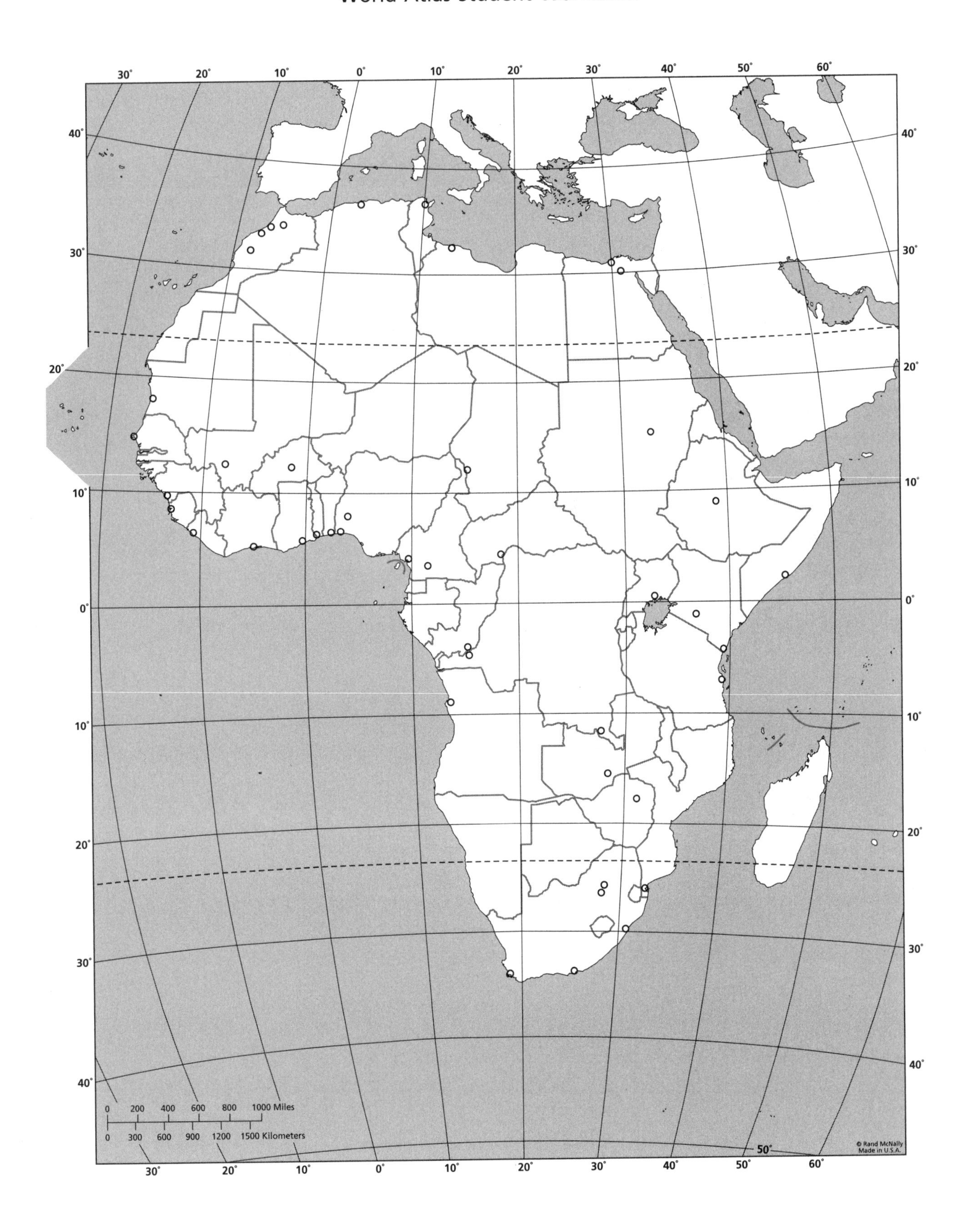
30°
20°
10°
0°
10°
20°
30°
40°
50°
60°
40°
30°
20°
10°
0°
10°
20°
30°
40°
0 200 400 600 800 1000 Miles
0 300 600 900 1200 1500 Kilometers
50°
© Rand McNally
Made in U.S.A.

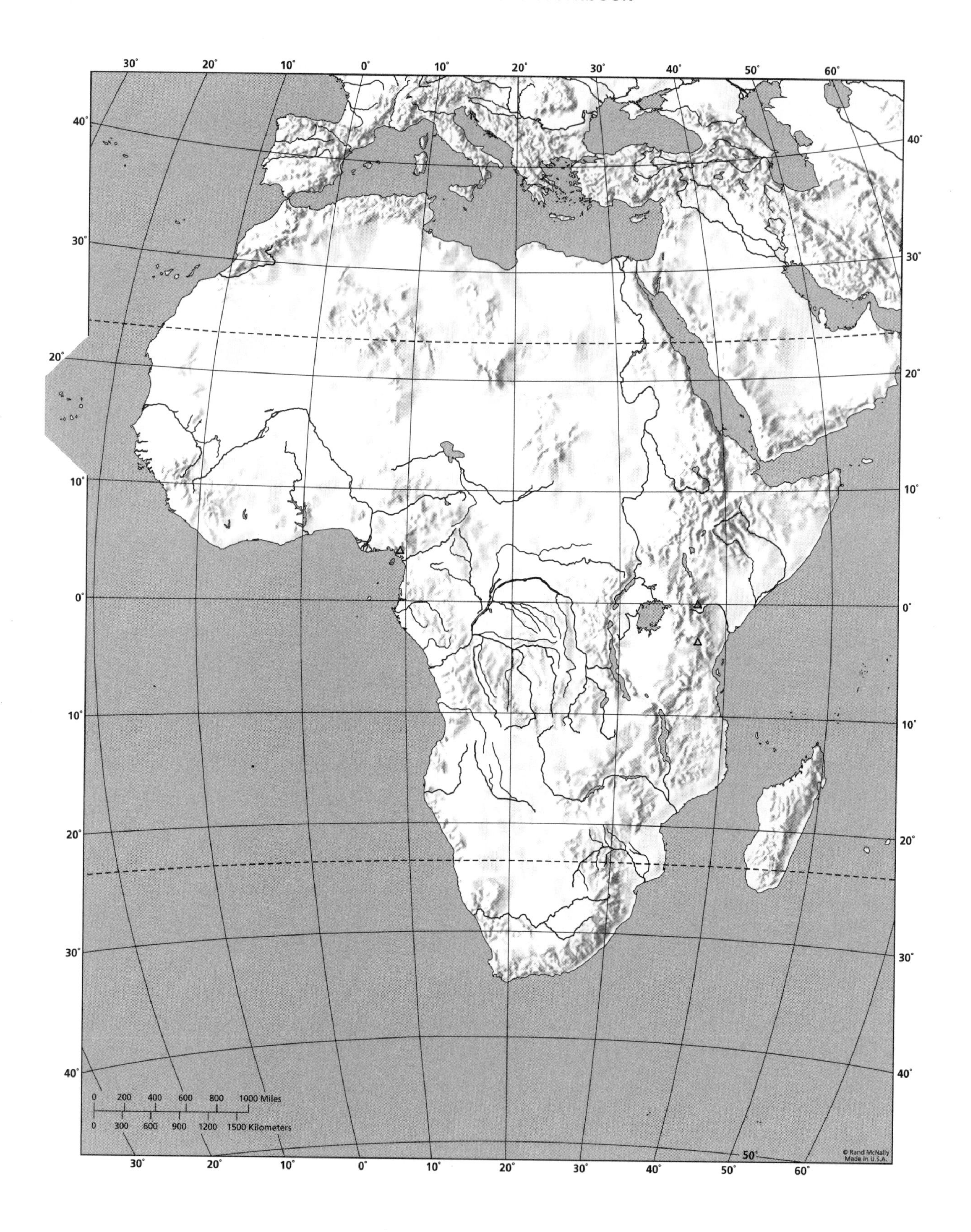
30°
20°
10°
0°
10°
20°
30°
40°
50°
60°
40°
30°
20°
10°
0°
10°
20°
30°
40°
0 200 400 600 800 1000 Miles
0 300 600 900 1200 1500 Kilometers
50°
© Rand McNally
Made in U.S.A.

# *Africa Blank Maps (Political / Physical)*

Name: ______________________ Date: ______________________

See pages **228-229** in Goode's World Atlas (21e) for full color maps.

## Political Map

Label the following **countries** on the blank political map of Africa.

| | | |
|---|---|---|
| Algeria<br>Angola<br>Benin<br>Botswana<br>Burkina Faso<br>Burundi<br>Cameroon<br>Central African Republic<br>Chad<br>Congo<br>Cote D'Ivoire<br>Democratic Republic of the Congo<br>Egypt<br>Equatorial Guinea<br>Eritrea | Ethiopia<br>Gabon<br>Ghana<br>Guinea<br>Guinea-Bissau<br>Kenya<br>Liberia<br>Libya<br>Madagascar<br>Malawi<br>Mali<br>Morocco<br>Mozambique<br>Namibia<br>Niger<br>Nigeria | Rwanda<br>Sao Tome and Principe<br>Senegal<br>Sierra Leone<br>Somalia<br>South Africa<br>Sudan<br>Tanzania<br>The Gambia<br>Togo<br>Tunisia<br>Uganda<br>Western Sahara<br>Zambia<br>Zimbabwe |

Label the following **cities** on the blank political map of Africa.

| | | |
|---|---|---|
| Abidjan<br>Accra<br>Addis Ababa<br>Alexandria<br>Algiers<br>Bamako<br>Bangui<br>Brazzaville<br>Cairo<br>Capetown<br>Casablanca<br>Conakry<br>Cotonou<br>Dakar<br>Dar es Salaam | Douala<br>Durban<br>Fès<br>Freetown<br>Harare<br>Ibadan<br>Johannesburg<br>Kampala<br>Khartoum<br>Kinshasa<br>Lagos<br>Lomé<br>Luanda<br>Lubumbashi<br>Lusaka | Maputo<br>Marrakech<br>Mogadishu<br>Mombasa<br>Monrovia<br>Nairobi<br>N'Djamena<br>Nouakchott<br>Ouagadougou<br>Port Elizabeth<br>Pretoria<br>Tripoli<br>Tunis<br>Yaoundé |

## Physical Map

Label the following **features** on the blank physical map of Africa.

### *Mountains & Mountain Ranges*

| | |
|---|---|
| Atlas<br>Cameroon Mt.<br>Drakensberg | Mount Kenya<br>Mount Kilimanjaro |

### *Rivers*

| | |
|---|---|
| Blue Nile<br>Congo<br>Limpopo<br>Niger | Nile<br>Orange<br>White Nile<br>Zambezi |

### *Bodies of Water*

| | |
|---|---|
| Gulf of Aden<br>Gulf of Guinea<br>Lake Albert<br>Lake Chad<br>Lake Edward<br>Lake Nyasa<br>Lake Rudolf | Lake Tana<br>Lake Tanganyika<br>Lake Victoria<br>Mediterranean Sea<br>Mozambique Channel<br>Red Sea |

Mauritius *Corbis Digital Stock*

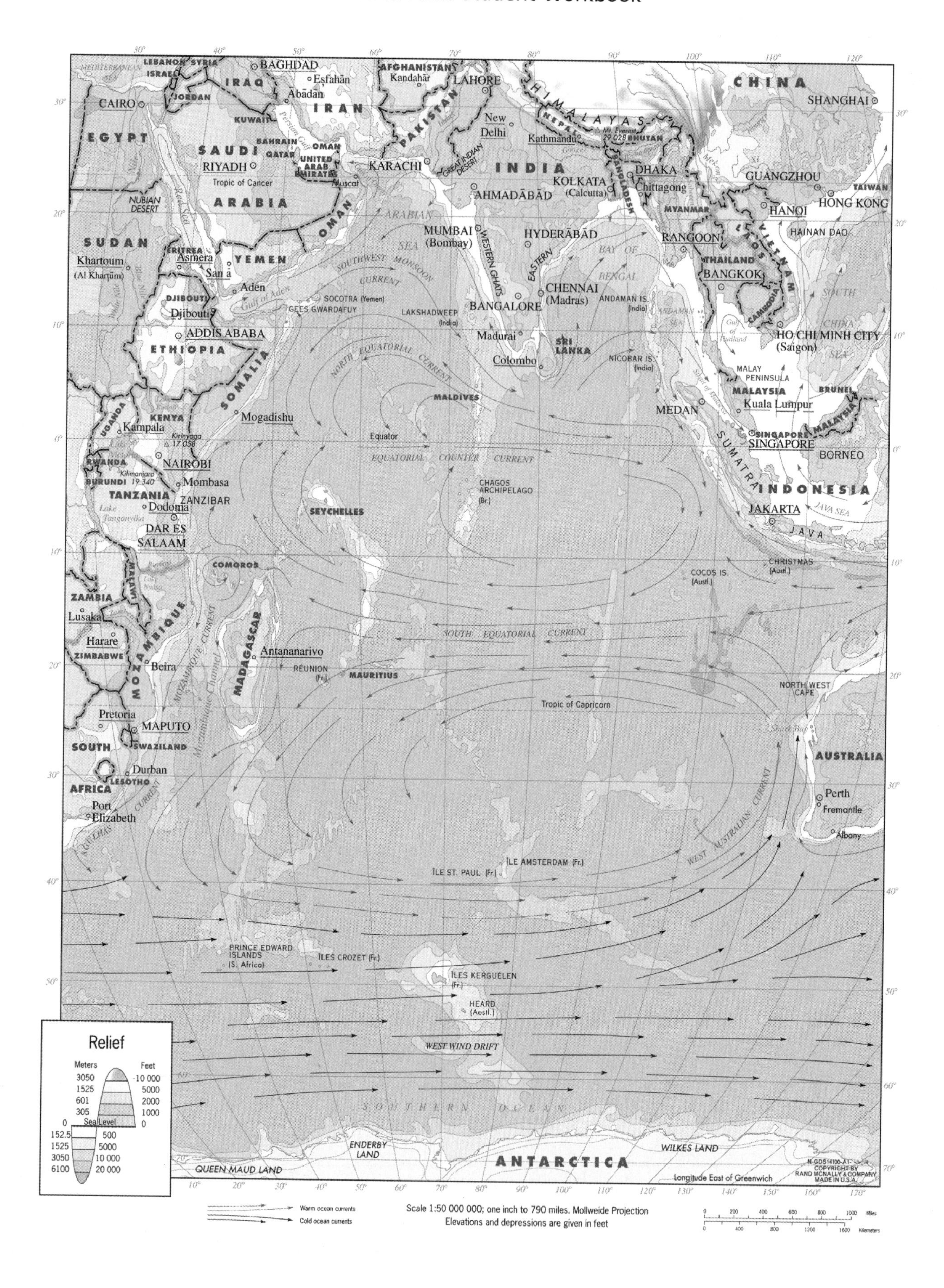
CHINA
SHANGHAI
INDIA
IRAN
IRAQ
SAUDI ARABIA
EGYPT
SUDAN
ETHIOPIA
SOMALIA
KENYA
TANZANIA
MADAGASCAR
MOZAMBIQUE
SOUTH AFRICA
AUSTRALIA
INDONESIA
ANTARCTICA
SOUTHERN OCEAN
ARABIAN SEA
BAY OF BENGAL
SOUTHWEST MONSOON CURRENT
NORTH EQUATORIAL CURRENT
EQUATORIAL COUNTER CURRENT
SOUTH EQUATORIAL CURRENT
WEST AUSTRALIAN CURRENT
WEST WIND DRIFT
AGULHAS CURRENT
MOZAMBIQUE CURRENT
Tropic of Cancer
Equator
Tropic of Capricorn
Relief
Meters
Feet
Sea Level
Warm ocean currents
Cold ocean currents
Scale 1:50 000 000; one inch to 790 miles. Mollweide Projection
Elevations and depressions are given in feet
Longitude East of Greenwich
COPYRIGHT BY RAND McNALLY & COMPANY MADE IN U.S.A.

## *Indian Ocean*

Name: ______________________ Date: ______________________

See page **239** in Goode's World Atlas (21e) for a full color map.

Describe the pattern of warm and cold ocean currents in the Indian Ocean.

What are the major island **groups** of the Indian Ocean?

______________________

______________________

______________________

______________________

______________________

______________________

What **countries** border the Indian Ocean?

Easter Island *Corbis Digital Stock*

Puna, Hawaii, USA *Corbis Digital Stock*

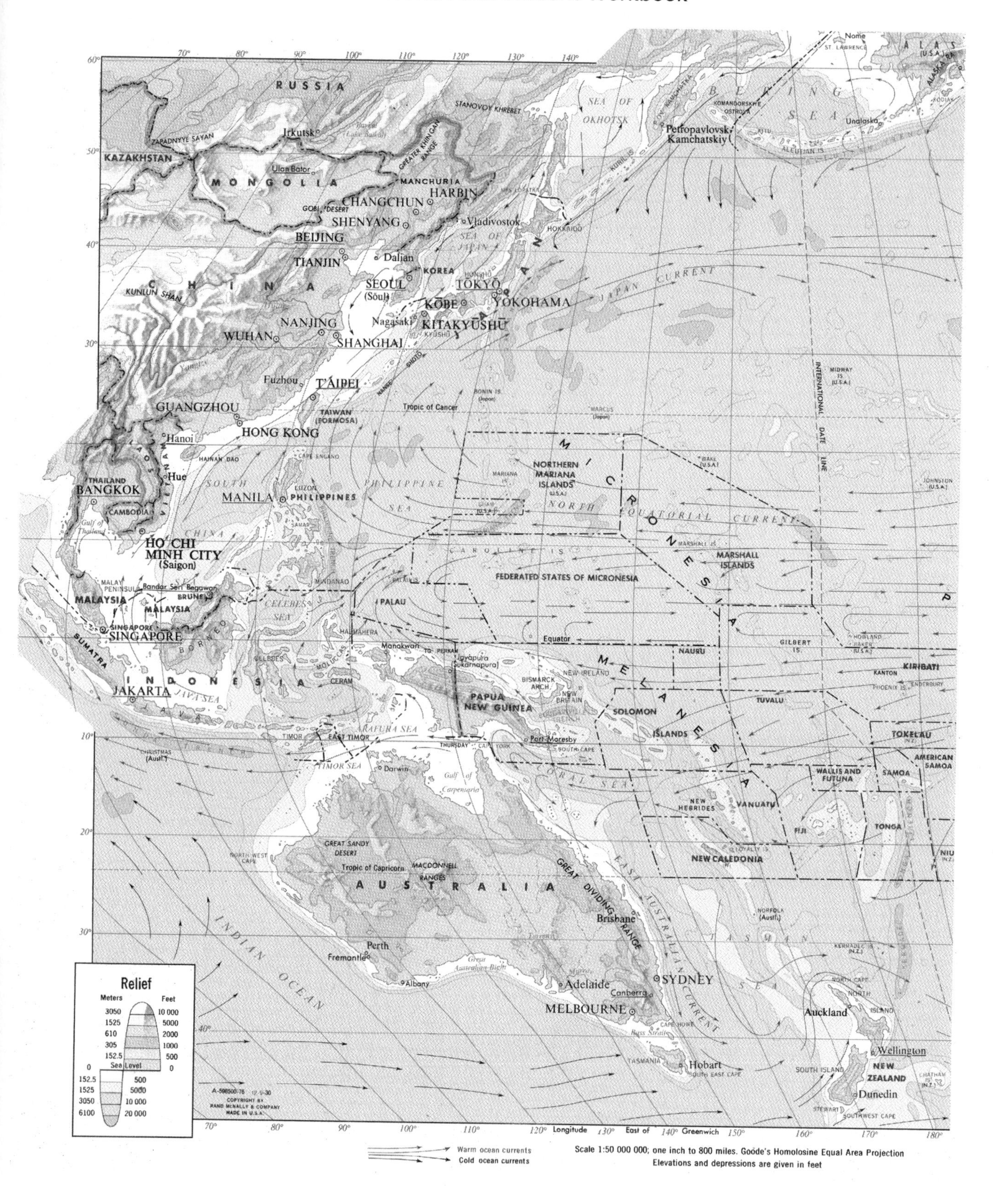


Warm ocean currents
Cold ocean currents

Scale 1:50 000 000; one inch to 800 miles. Goode's Homolosine Equal Area Projection
Elevations and depressions are given in feet

a

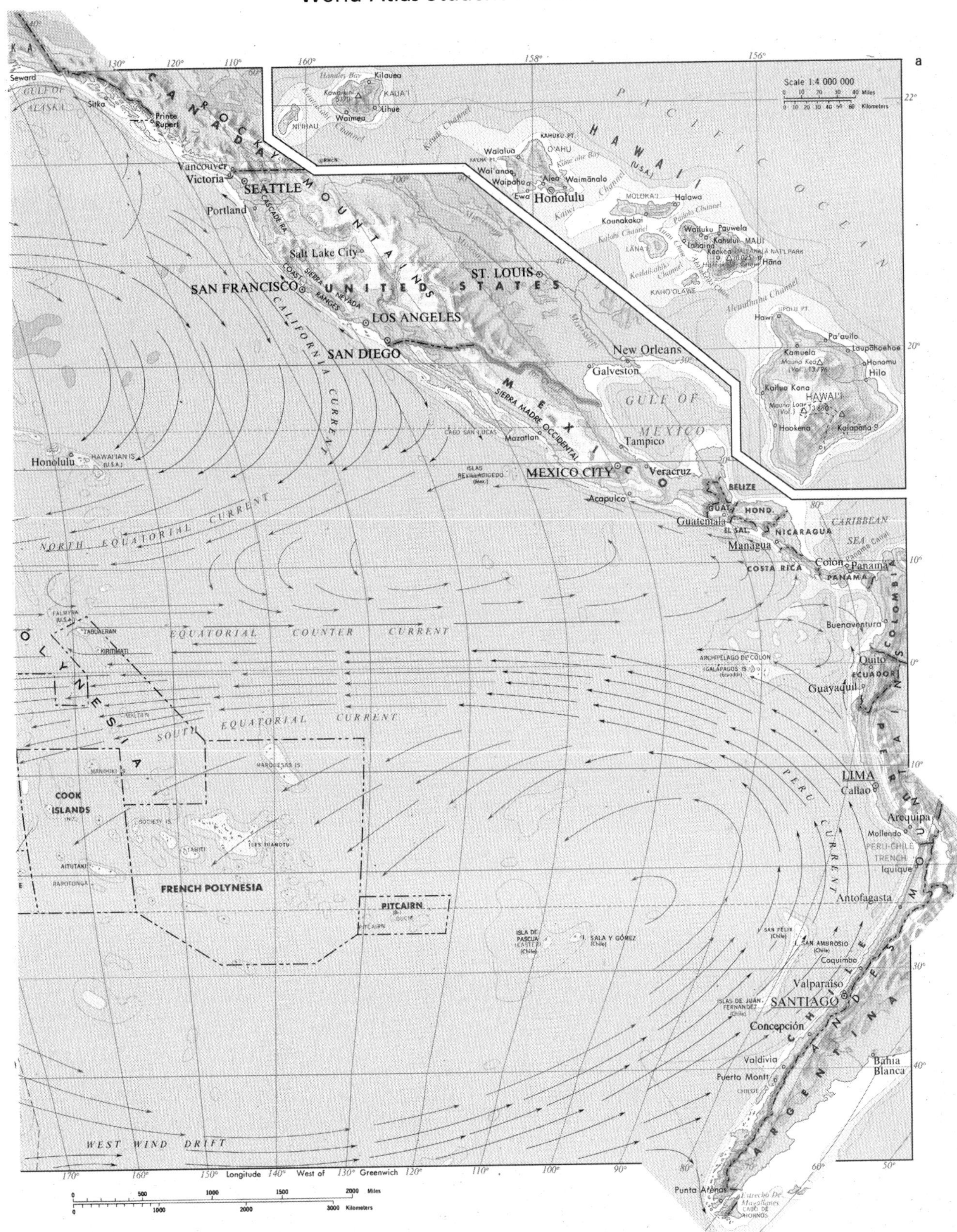

PACIFIC OCEAN
HAWAII
Honolulu
ROCKY MOUNTAINS
CANADA
UNITED STATES
SEATTLE
Portland
Salt Lake City
SAN FRANCISCO
LOS ANGELES
SAN DIEGO
ST. LOUIS
New Orleans
Galveston
GULF OF MEXICO
Tampico
MEXICO
MEXICO CITY
Veracruz
Acapulco
Mazatlan
CALIFORNIA CURRENT
NORTH EQUATORIAL CURRENT
EQUATORIAL COUNTER CURRENT
SOUTH EQUATORIAL CURRENT
PERU CURRENT
WEST WIND DRIFT
CARIBBEAN SEA
Guatemala
Managua
Colón
Panama
COSTA RICA
Buenaventura
Quito
ECUADOR
Guayaquil
LIMA
Callao
Arequipa
Antofagasta
Valparaíso
SANTIAGO
Concepción
Valdivia
Puerto Montt
Punta Arenas
Bahía Blanca
COOK ISLANDS
FRENCH POLYNESIA
PITCAIRN
POLYNESIA
Vancouver
Victoria
Sitka
Seward

Tahiti *Corbis Digital Stock*

## *Pacific Ocean*

Name: ______________________________ Date: ______________________________

See pages **240-241** in Goode's World Atlas (21e) for a full color map.

Describe the pattern of warm and cold ocean currents in the Indian Ocean.

What are the major **islands** and **island groups** of Melanesia?

______________________________________________________________________
______________________________________________________________________
______________________________________________________________________
______________________________________________________________________
______________________________________________________________________
______________________________________________________________________

What are the major **islands** and **island groups** of Micronesia?

______________________________________________________________________
______________________________________________________________________
______________________________________________________________________
______________________________________________________________________
______________________________________________________________________
______________________________________________________________________

What are the major **islands** and **island groups** of Polynesia?

What are the northernmost islands of the Pacific?

What are the southernmost islands of the Pacific?

What countries border the Pacific?

Where are the deepest trenches in the Pacific?

Rockport, Maine, USA *Corbis Digital Stock*

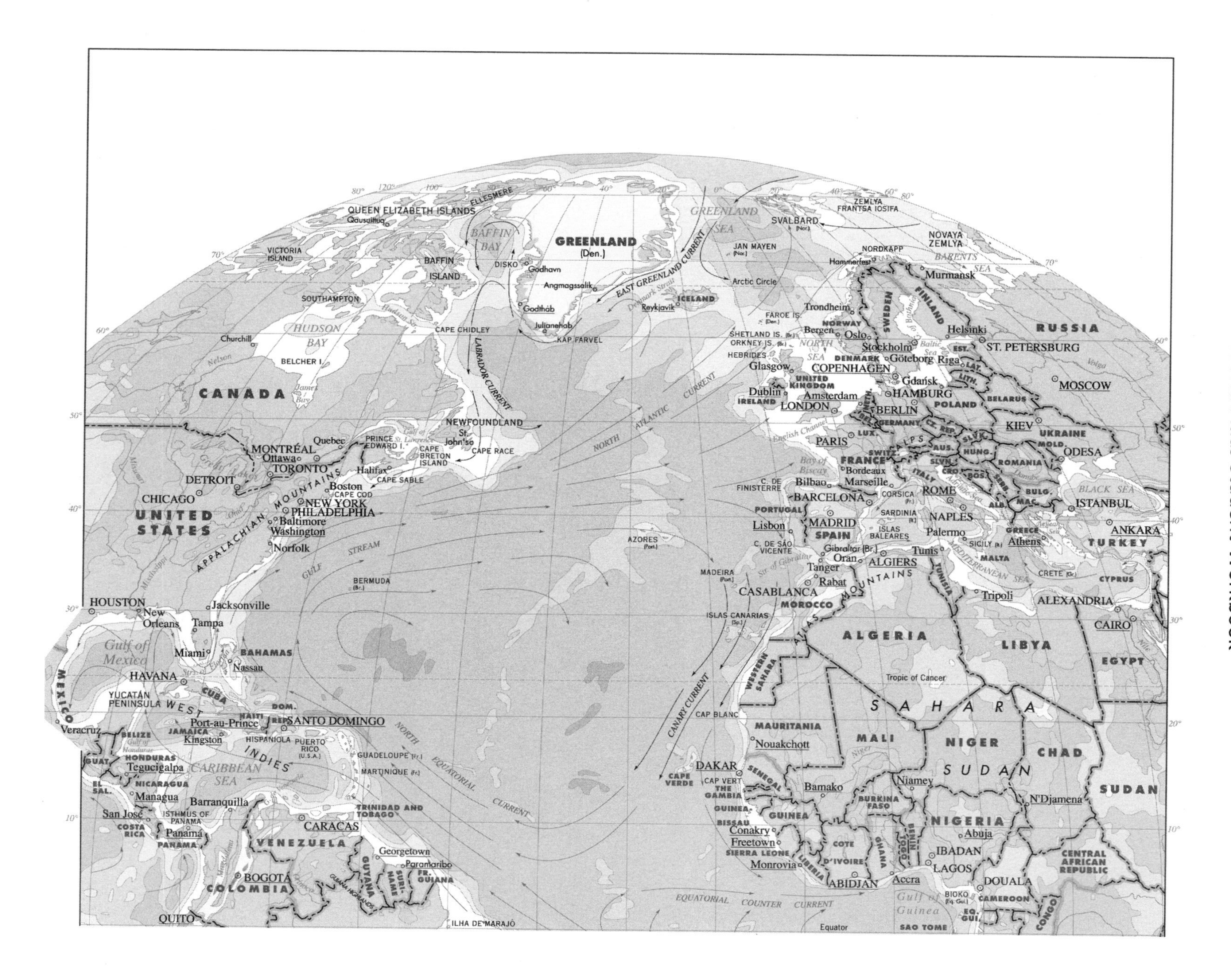

QUEEN ELIZABETH ISLANDS
ELLESMERE
BAFFIN BAY
GREENLAND (Den.)
GREENLAND SEA
SVALBARD
ZEMLYA FRANTSA IOSIFA
NOVAYA ZEMLYA
BARENTS SEA
JAN MAYEN
NORDKAPP
Hammerfest
Murmansk
VICTORIA ISLAND
BAFFIN ISLAND
DISKO
Godhavn
Angmagssalik
EAST GREENLAND CURRENT
Arctic Circle
ICELAND
Reykjavik
SOUTHAMPTON
Godthåb
Julianehåb
KAP FARVEL
CAPE CHIDLEY
HUDSON BAY
Churchill
BELCHER I.
CANADA
LABRADOR CURRENT
NEWFOUNDLAND
St. John's
CAPE RACE
PRINCE EDWARD I.
CAPE BRETON ISLAND
Halifax
CAPE SABLE
MONTRÉAL
Quebec
Ottawa
TORONTO
DETROIT
CHICAGO
UNITED STATES
APPALACHIAN MOUNTAINS
Boston
CAPE COD
NEW YORK
PHILADELPHIA
Baltimore
Washington
Norfolk
GULF STREAM
BERMUDA (Br.)
HOUSTON
New Orleans
Jacksonville
Tampa
Miami
Gulf of Mexico
MEXICO
BAHAMAS
Nassau
HAVANA
CUBA
YUCATÁN PENINSULA
WEST INDIES
DOM. REP.
HAITI
SANTO DOMINGO
Port-au-Prince
JAMAICA
Kingston
HISPANIOLA
PUERTO RICO (U.S.A.)
GUADELOUPE (Fr.)
MARTINIQUE (Fr.)
Veracruz
BELIZE
GUAT.
HONDURAS
Tegucigalpa
CARIBBEAN SEA
EL SAL.
NICARAGUA
Managua
Barranquilla
San José
COSTA RICA
ISTHMUS OF PANAMA
Panamá
PANAMA
TRINIDAD AND TOBAGO
CARACAS
VENEZUELA
Georgetown
Paramaribo
GUYANA
SURINAME
FR. GUIANA
BOGOTÁ
COLOMBIA
QUITO
ILHA DE MARAJÓ
NORTH EQUATORIAL CURRENT
NORTH ATLANTIC CURRENT
AZORES (Port.)
MADEIRA (Port.)
ISLAS CANARIAS (Sp.)
CANARY CURRENT
CAP BLANC
CAPE VERDE
DAKAR
EQUATORIAL COUNTER CURRENT
Equator
FAROE IS. (Den.)
SHETLAND IS. (Br.)
ORKNEY IS. (Br.)
HEBRIDES
Trondheim
NORWAY
Bergen
Oslo
SWEDEN
FINLAND
Helsinki
RUSSIA
ST. PETERSBURG
MOSCOW
Stockholm
Göteborg
Riga
EST.
LAT.
LITH.
DENMARK
COPENHAGEN
NORTH SEA
Glasgow
UNITED KINGDOM
Dublin
IRELAND
LONDON
Amsterdam
Gdańsk
HAMBURG
BERLIN
POLAND
BELARUS
GERMANY
KIEV
UKRAINE
MOLD.
ODESA
PARIS
LUX.
SWITZ.
FRANCE
Bordeaux
Marseille
ALPS
AUS.
HUNG.
ROMANIA
SLVN.
CRO.
BOS.
SERB.
ITALY
Bay of Biscay
English Channel
C. DE FINISTERRE
Bilbao
BARCELONA
CORSICA (Fr.)
ROME
BULG.
MAC.
ALB.
BLACK SEA
ISTANBUL
PORTUGAL
Lisbon
MADRID
SPAIN
SARDINIA (It.)
ISLAS BALEARES
NAPLES
Palermo
SICILY (It.)
GREECE
Athens
ANKARA
TURKEY
C. DE SÃO VICENTE
Gibraltar (Br.)
Oran
ALGIERS
Tunis
MALTA
MEDITERRANEAN SEA
CRETE (Gr.)
CYPRUS
Tanger
Rabat
ATLAS MOUNTAINS
CASABLANCA
MOROCCO
TUNISIA
Tripoli
ALEXANDRIA
CAIRO
ALGERIA
LIBYA
EGYPT
WESTERN SAHARA
Tropic of Cancer
SAHARA
MAURITANIA
Nouakchott
MALI
NIGER
CHAD
SUDAN
SENEGAL
CAP VERT
THE GAMBIA
GUINEA-BISSAU
GUINEA
Conakry
Freetown
SIERRA LEONE
Monrovia
LIBERIA
Bamako
BURKINA FASO
Niamey
N'Djamena
NIGERIA
Abuja
IBADAN
LAGOS
COTE D'IVOIRE
ABIDJAN
GHANA
Accra
TOGO
BENIN
Gulf of Guinea
BIOKO (Eq. Gui.)
DOUALA
CAMEROON
EQ. GUI.
SAO TOME
CENTRAL AFRICAN REPUBLIC
CONGO

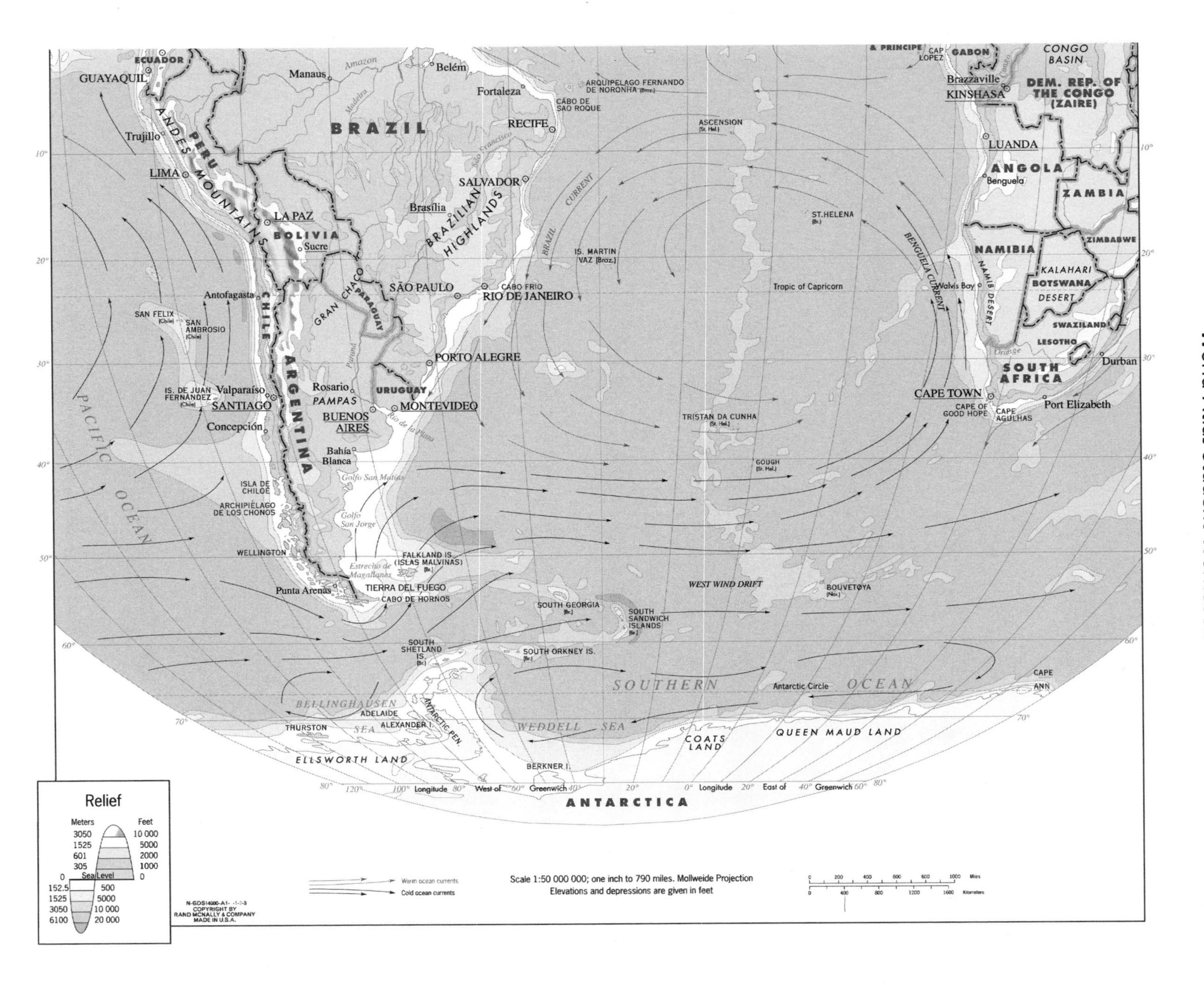

ECUADOR
GUAYAQUIL
Manaus
Amazon
Belém
Fortaleza
ARQUIPÉLAGO FERNANDO DE NORONHA
CABO DE SÃO ROQUE
RECIFE
BRAZIL
Trujillo
PERU
ANDES MOUNTAINS
LIMA
SALVADOR
BRAZILIAN HIGHLANDS
Brasília
LA PAZ
BOLIVIA
Sucre
IS. MARTIN VAZ
BRAZIL CURRENT
ASCENSION
ST. HELENA
Tropic of Capricorn
SÃO PAULO
CABO FRIO
RIO DE JANEIRO
GRAN CHACO
PARAGUAY
Antofagasta
CHILE
SAN FÉLIX
SAN AMBROSIO
ARGENTINA
PORTO ALEGRE
IS. DE JUAN FERNÁNDEZ
Valparaíso
SANTIAGO
Concepción
Rosario
PAMPAS
URUGUAY
MONTEVIDEO
BUENOS AIRES
Bahía Blanca
PACIFIC OCEAN
ISLA DE CHILOÉ
ARCHIPIÉLAGO DE LOS CHONOS
WELLINGTON
FALKLAND IS. (ISLAS MALVINAS)
Punta Arenas
TIERRA DEL FUEGO
CABO DE HORNOS
SOUTH GEORGIA
SOUTH SANDWICH ISLANDS
WEST WIND DRIFT
BOUVETØYA
TRISTAN DA CUNHA
GOUGH
SOUTH SHETLAND IS.
SOUTH ORKNEY IS.
SOUTHERN OCEAN
Antarctic Circle
BELLINGSHAUSEN SEA
ADELAIDE
ALEXANDER I.
ANTARCTIC PEN.
THURSTON
ELLSWORTH LAND
WEDDELL SEA
BERKNER I.
COATS LAND
QUEEN MAUD LAND
CAPE ANN
ANTARCTICA
CONGO BASIN
GABON
CAP LOPEZ
Brazzaville
KINSHASA
DEM. REP. OF THE CONGO (ZAIRE)
LUANDA
ANGOLA
Benguela
ZAMBIA
ZIMBABWE
NAMIBIA
KALAHARI
BOTSWANA
DESERT
NAMIB DESERT
Walvis Bay
BENGUELA CURRENT
SWAZILAND
LESOTHO
SOUTH AFRICA
Durban
CAPE TOWN
CAPE OF GOOD HOPE
CAPE AGULHAS
Port Elizabeth
Relief
Meters
Feet
Sea Level
Warm ocean currents
Cold ocean currents
Scale 1:50 000 000; one inch to 790 miles. Mollweide Projection
Elevations and depressions are given in feet
COPYRIGHT BY RAND McNALLY & COMPANY
MADE IN U.S.A.

Virgin Gorda, British Virgin Islands *Digital Vision*

## *Atlantic Ocean*

Name: ______________________________ Date: ___________________________

See pages **242-243** in Goode's World Atlas (21e) for a full color map.

Describe the pattern of warm and cold ocean currents in the Atlantic Ocean.

What are the major **islands** and **island groups** of the Atlantic north of the Equator?

______________________________________________________________________

______________________________________________________________________

______________________________________________________________________

______________________________________________________________________

______________________________________________________________________

______________________________________________________________________

What are the major **islands** and **island groups** of the Atlantic south of the Equator?

______________________________________________________________________

______________________________________________________________________

______________________________________________________________________

______________________________________________________________________

______________________________________________________________________

______________________________________________________________________

What countries border the Atlantic Ocean?

Can you make any connections between historical trade patterns among Europe, Africa, and the Americas and the direction of the ocean currents in the North Atlantic?

Churchill, Manitoba, Canada *Corbis Digital Stock*

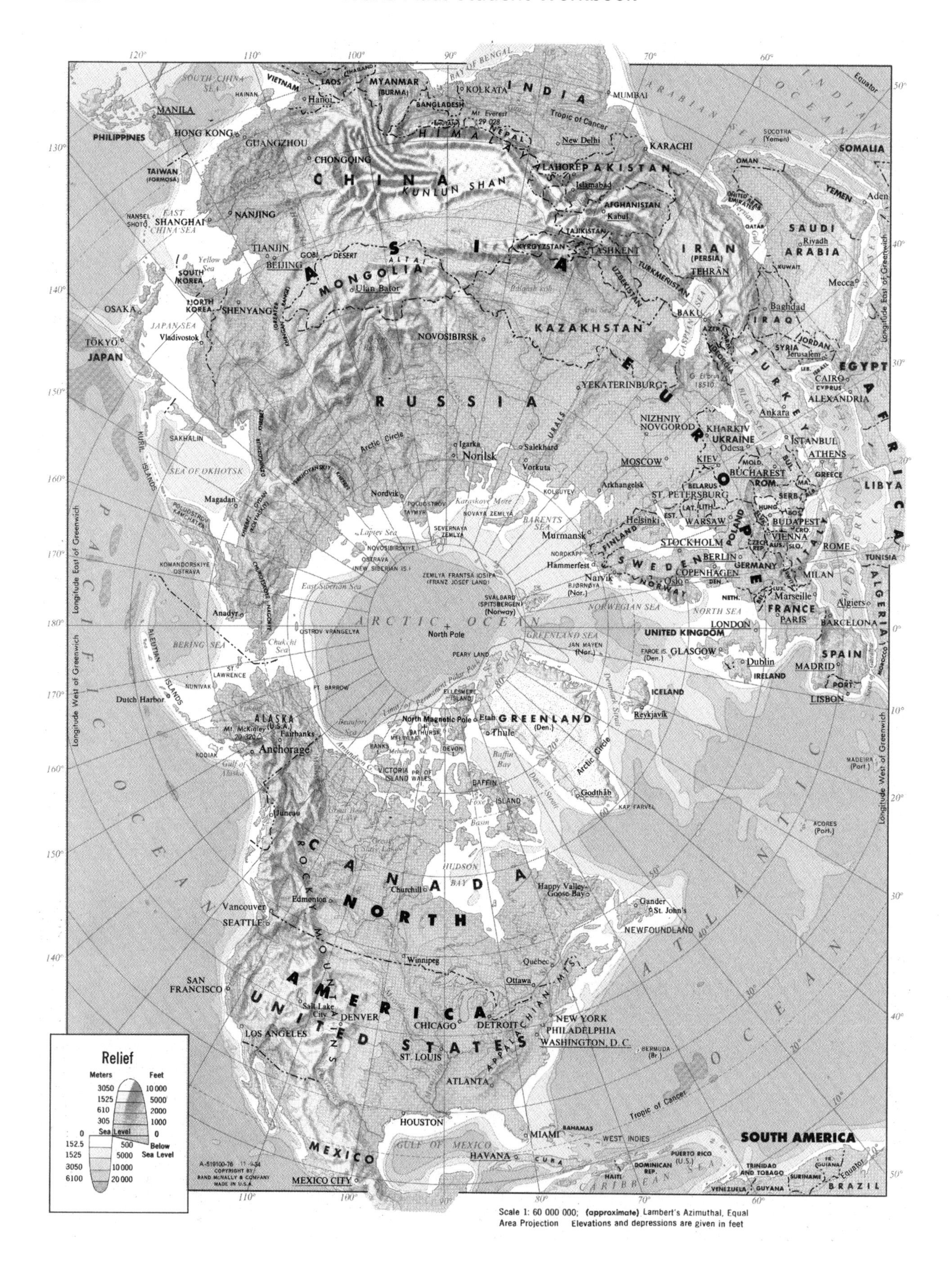


Scale 1: 60 000 000; (approximate) Lambert's Azimuthal, Equal Area Projection Elevations and depressions are given in feet

## *Arctic Ocean*

Name: ______________________ Date: ______________________

See page **244** in Goode's World Atlas (21e) for a full color map.

What are the major **seas** of the Arctic Ocean?

What are the major **islands** of the Arctic Ocean?

What countries border the Arctic Ocean?